William Dawson

The Origin of the World According to Revelation and Science

William Dawson

The Origin of the World According to Revelation and Science

ISBN/EAN: 9783744708418

Printed in Europe, USA, Canada, Australia, Japan

Cover: Foto ©berggeist007 / pixelio.de

More available books at **www.hansebooks.com**

William Dawson

The Origin of the World According to Revelation and Science

ISBN/EAN: 9783744708418

Printed in Europe, USA, Canada, Australia, Japan

Cover: Foto ©berggeist007 / pixelio.de

More available books at **www.hansebooks.com**

THE

ORIGIN OF THE WORLD,

ACCORDING TO

REVELATION AND SCIENCE.

By J. W. DAWSON, LL.D., F.R.S., F.G.S.,

PRINCIPAL AND VICE-CHANCELLOR OF M'GILL UNIVERSITY, MONTREAL; AUTHOR OF
"ACADIAN GEOLOGY," "THE STORY OF THE EARTH AND MAN,"
"LIFE'S DAWN ON EARTH," ETC.

"Speak to the Earth, and it shall teach thee."
—*Job.*

MONTREAL:
DAWSON BROTHERS, PUBLISHERS,
159-161 ST. JAMES STREET.
1877.

TO HIS EXCELLENCY

THE RIGHT HON. THE EARL OF DUFFERIN,
K.P., K.C.B., ETC.,

GOVERNOR-GENERAL OF CANADA,

𝕿𝖍𝖎𝖘 𝖂𝖔𝖗𝖐 𝖎𝖘 𝕽𝖊𝖘𝖕𝖊𝖈𝖙𝖋𝖚𝖑𝖑𝖞 𝕯𝖊𝖉𝖎𝖈𝖆𝖙𝖊𝖉,

AS A SLIGHT TRIBUTE OF ESTEEM TO ONE WHO GRACES THE
HIGHEST POSITION IN THE DOMINION OF CANADA BY HIS
EMINENT PERSONAL QUALITIES, HIS REPUTATION AS
A STATESMAN AND AN AUTHOR, AND HIS KIND
AND ENLIGHTENED PATRONAGE OF EDU-
CATION, LITERATURE, AND SCIENCE.

PREFACE.

THE scope of this work is in the main identical with that of " Archaia," published in 1860; but in attempting to prepare a new edition brought up to the present condition of the subject, it was found that so much required to be rewritten as to make it essentially a new book, and it was therefore decided to give it a new name, more clearly indicating its character and purpose.

The intention of this new publication is to throw as much light as possible on the present condition of the much-agitated questions respecting the origin of the world and its inhabitants. To students of the Bible it will afford the means of determining the precise import of the biblical references to creation, and of their relation to what is known from other sources. To geologists and biologists it is intended to give some intelligible explanation of the connection of the doctrines of revealed religion with the results of their respective sciences.

A still higher end to which the author would gladly contribute is that of aiding thoughtful men perplexed with the apparent antagonisms of science and religion, and of indicating how they may best harmonize our great and growing knowledge of nature with our old and cherished beliefs as to the origin and destiny of man.

In aiming at these results, it has not been thought

A

necessary to assume a controversial attitude or to stand on the defensive, either with regard to religion or science, but rather to attempt to arrive at broad and comprehensive views which may exhibit those higher harmonies of the spiritual and the natural which they derive from their common Author, and which reach beyond the petty difficulties arising from narrow or imperfect views of either or both. Such an aim is too high to be fully attained, but in so far as it can be reached we may hope to rescue science from a dry and barren infidelity, and religion from mere fruitless sentiment or enfeebling superstition.

Since the publication of "Archaia," the subject of which it treats has passed through several phases, but the author has seen no reason to abandon in the least degree the principles of interpretation on which he then insisted, and he takes a hopeful view as to their ultimate prevalence. It is true that the wide acceptance of hypotheses of "evolution" has led to a more decided antagonism than heretofore between some of the utterances of scientific men and the religious ideas of mankind, and to a contemptuous disregard of revealed religion in the more shallow literature of the time; but, on the other hand, a barrier of scientific fact and induction has been slowly rising to stem this current of crude and rash hypothesis. Of this nature are the great discoveries as to the physical constitution and probable origin of the universe, the doctrine of the correlation and conservation of forces, the new estimates of the age of the earth, the overthrow of the doctrine of spontaneous generation, the high bodily and mental type of the earliest known men, the light which philology has thrown on the unity of language, our growing knowledge of the uniformity of the constructive and other habits of primitive men, and of the condition of man in the earlier historic time, the greater completeness of our conceptions as to the phenomena of life and their relation to organizable matters — all these and many other aspects of the later progress of science must tend to bring it back into greater harmony with revealed religion.

On the other side, there has been a growing disposition on the part of theologians to inquire as to the actual views of nature presented in the Bible, and to separate these from those accretions of obsolete philosophy which have been too often confounded with them.   With respect to the first chapter of Genesis more especially, there has been a decided growth in the acceptance of those principles for which I contended in 1860.   In illustration of this I may refer to the fact that in 1862 it was precisely on these principles that Dr. McCaul conducted his able defence of the Mosaic record of creation in the "Aids to Faith," which may almost be regarded as an authoritative expression of the views of orthodox Christians in opposition to those of the once notorious "Essays and Reviews." Equally significant is the adoption of this method of interpretation by Dr. Tayler Lewis in his masterly "Special Introduction" to the first chapter of Genesis, in the American edition of Lange's Commentary, edited by Dr. Philip Schaff; and the manifest approval with which the lucid statement of the relations of Geology and the Bible by Dr. Arnold Guyot, was received by the great gathering of divines at the Convention of the Evangelical Alliance in New York, in 1873, bears testimony to the same fact.   The author has also had the honor of being invited to illustrate this mode of reconciliation to the students of two of the most important theological colleges in America, in lectures afterwards published and widely circulated.

The time is perhaps nearer than we anticipate when Natural Science and Theology will unite in the conviction that the first chapter of Genesis " stands alone among the traditions of mankind in the wonderful simplicity and grandeur of its words," and that " the meaning of these words is always a meaning ahead of science—not because it anticipates the results of science, but because it is independent of them, and runs as it were round the outer margin of all possible discovery."*

In the Appendix the reader will find several short essays on special points collateral to the general subject, and important in the solution of some of its difficulties, but which could not be conveniently included in the text. More especially I would refer to the summaries given in the Appendix of the present state of our knowledge as to the origin of life, of species, and of man—topics not discussed in much detail in the body of the work, both because of the wide fields of controversy to which they lead, and because I have treated of them somewhat fully in a previous work, " The Story of the Earth and Man," in which the detailed history of life as disclosed by science was the main subject in hand.

J. W. D.

May, 1877.

---

* Argyll's " Primeval Man."

CONTENTS.

CHAPTER VI.

LIGHT AND CREATIVE DAYS.

CHAPTER VII.

THE ATMOSPHERE.

CHAPTER VIII.

THE DRY LAND AND THE FIRST PLANTS.

CHAPTER IX.

LUMINARIES.

CHAPTER X.

THE LOWER ANIMALS.

CHAPTER XI.

THE HIGHER ANIMALS AND MAN.

CHAPTER XII.

THE REST OF THE CREATOR.

CHAPTER XIII.

UNITY AND ANTIQUITY OF MAN.

CHAPTER XIV.

UNITY AND ANTIQUITY OF MAN (*continued*).

CHAPTER XV.

COMPARISONS AND CONCLUSIONS.

APPENDICES.

THE ORIGIN OF THE WORLD.

CHAPTER I.

THE MYSTERY OF ORIGINS AND ITS SOLUTIONS.

"The things that are seen are temporal."—PAUL.

HAVE we or can we have any certain solution of those two great questions—Whence are all things? and Whither do all things tend? No thinking man is content to live merely in a transitory present, ever emerging out of darkness and ever returning thither again, without knowing any thing of the origin and issue of the world and its inhabitants. Yet it would seem that to-day men are as much in uncertainty on these subjects as at any previous time. It even appears as if all our added knowledge would only, for a time at least, deprive us of the solutions to which we trusted, and give no others in their room. Christians have been accustomed to rest on the cosmogony and prophecy of the Bible; but we are now frankly told on all hands that these are valueless, and that even ministers of religion more or less "sacrifice their sincerity" in making them the basis of their teachings. On the other hand, we are informed that nothing can be discerned in the universe beyond matter and force, and that it is by a purely material and spontaneous evolution that all things

A 2

exist.   But when we ask as to the origin of matter and force, and the laws which regulate them—as to the end to which their movement is tending, as to the manner in which they have evolved the myriad forms of life and the human intelligence itself—the only answer is that these are "insoluble mysteries."

Are we, then, to fall back on the real or imagined revelations and traditions of the past, and to endeavor to find in them some foothold of assurance; or are we to wait till further progress in science may have cleared up some of the present mysteries?   Whatever may be said of the former alternative, all honest students of science will unite with me in the admission that the latter is hopeless.   We need not seek to belittle the magnificent triumphs of modern science.   They have been real and stupendous.   But it is of their very nature to conduct us to ultimate facts and laws of which science can give no explanation; and the further we push our inquiries the more insuperably does the wall of mystery rise before us. It is true we can furnish the materials for philosophical speculations which may be built on scientific facts and principles; but these are in their nature uncertain, and must constantly change as knowledge advances.   They can not solve for us the great practical problems of our origin and destiny.

In these circumstances no apology is needed for a thorough and careful inquiry into those foundations of religious belief which rest on the idea of a revelation of origins and destinies made to man from without, and on which we may build the superstructure of a rational religion, giving guidance for the present and hope for the future.   In the following pages I propose to enter upon so much of this subject as relates to the origin and earliest history of the world, in so far as these are treated of in the Bible and in the traditions of the more an-

cient nations ; and this with reference to the present stand-
point of science in relation to these questions.

To discuss such questions at all, certain preliminary ad-
missions are necessary. These are : (1) The reality of an unseen universe, spiritual rather than material in its nature.
(2) The existence of a personal God, or of a great Universal
Will.   (3) The possibility of communication taking place between God and man.   I do not propose to attempt any proof
of these positions, but it may be well to explain what they
mean.

(1) That the great machine for the dissipation of energy, in
which we exist, and which we call the universe, must have a
correlative and complement in the unseen, is a conclusion
now forced upon physicists by the necessities of the doctrine
of the conservation of force.   In short, it seems that, unless
we admit this conclusion, we can not believe in the possible
existence of the material universe itself, and must sink into
absolute nihilism.   This doctrine is expressed by the apostle
Paul in the statement, " The things that are seen are temporal, but the things that are not seen are eternal," and it has
been ably discussed by the authors of the remarkable work,
" The Unseen Universe."   That this unseen world is spiritual—that is, not subject to the same material laws with the
visible universe—is also a fair deduction from physical science,
as well as a doctrine of Scripture.   I prefer the term spiritual
to supernatural, because the first is the term used in the Bible,
and because the latter has had associated with it ideas of the
miraculous and abnormal, not implied at all in the idea of
the spiritual, which in some important senses may be more
natural than the material.

(2) The idea of a personal God implies not merely the existence of an unknown absolute power, as Herbert Spencer

seems to hold, or of "an Eternal, not ourselves, that makes for righteousness," as Matthew Arnold puts it, but of a Being of whom we can affirm will, intelligence, feeling, self-consciousness, not certainly precisely as they occur in us, but in a higher and more perfect form, of which our own consciousness furnishes the type, or "image and shadow," as Moses long ago phrased it.   On the one hand, it is true that we can not fully comprehend such a personal God, because not limited by the conditions which limit us.   On the other hand, it is clear that our intellect, as constituted, can furnish us with no ultimate explanation of the universe except in the action of such a primary personal will.   In the Bible the absolute personality of God is expressed by the title "I am."   His intimate relation to us is indicated by the expression, "In him we live, and move, and have our being."   His all-pervading essence is stated as "the fullness of him that filleth all in all."   His relative personality is shadowed forth by the attribution to him of love, anger, and other human feelings and sentiments, and by presenting him in the endearing relation of the universal Father.

(3) With reference to the possibility of communication between God and man, it may truly be said that such communication is not only possible, but infinitely probable.   God is not only near to us, but we are in him, and, independently of the testimony of revelation, it has been felt by all classes of men, from the rudest and most primitive savages up to our great English philosopher, John Stuart Mill, that if there is a God, he can not be excluded from communion with his intelligent creatures, either directly or through the medium of ministering spirits.*   Farther, placed as man is in the midst

* Essays on Theism, 1875.

of complex and to him inexplicable phenomena, involved in a conflict of good and evil, happiness and misery, to which the wisest and the greatest minds have found no issue, subject to be degraded by low passions and tempted to great extremes of evil, and himself weak, impulsive, and vacillating, there seems the most urgent need for divine communication. It may be said that these are conflicts and problems which God has left man to decide and solve for himself by his own reason. But when we consider how slow this process is, and how imperfect even now, after the experience of ages, we seem to need some intervention that shall stimulate the human mind, and impel it forward with greater rapidity. Farther, it would appear only right that an intelligent and accountable being, placed in a world like this, should have some explanation of his origin and destiny given him at first, and that, if he should perchance gó astray, a helping hand should be extended to him.

Practically it is an historical fact that all the great impulses given to humanity have been by men claiming divine guidance or inspiration, and professing to bring light and truth from the unseen world. It would be too much to say that all these prophets and reformers have been inspired of heaven ; but scarcely too much to say that they have either received a message of God, or have been permitted to transmit to our world messages for weal or woe from powers without in subordination to him. Farther, we shall have reason in the sequel to see that in far back prehistoric times there must have been impulses given to mankind, and revelations made to them, as potent as those which have acted in later historic periods. In Holy Scripture the Word of God is represented as "enlightening every man ;* and with reference to

---

* John i., 9.

our present subject we are told that " by faith we understand
that the ages of the world were constituted by the Word of
God, so that the visible things were not made of those which
appear."*   In other words, that the will of God has been active and operative as the sole cause throughout all ages of the
world's creation and history, and that the visible universe is
not a mere product of Its own phenomena.   We may call
this faith, if we please, an intuition or instinct, a God-given
gift, or a product of our own thought acting on evidence afforded by the outer world ; but in any case it seems to be
the sole possible solution of the mystery of origins.

These points being premised, we are in a position to inquire as to the teaching of our own Holy Scriptures, and in
this inquiry we can easily take along with them all other revelations, pretended or true, that deal with our subject.

Max Müller, in his lectures on the Science of Religion, rejects the ordinary division into natural and revealed, and
adopts a threefold grouping, corresponding to the great
division of languages into Turanian, Aryan, and Semitic.
With some modification and explanation, this classification
will serve well our present purpose.   As to natural and revealed religions, if we regard our own as revealed, we must
admit an element of revelation in all others as well.   According to the Hebrew Scriptures revelation began in Eden,
and was continued more or less in all successive ages up to
the apostolic times.   Consequently the earlier revelations of
the antediluvian and postdiluvian times must have been the
common property of all races, and must have been associated
with whatever elements of natural religion they had.   When,
therefore, we call our religion distinctively a revealed one, we

* Hebrews xi., 3.

must admit that traces of the same revelation may be found in all others. On the other hand, when we characterize our religion as Hebrew or Semitic, we must bear in mind that in its earlier stages it was not so limited ; but that, if as old as it professes to be, it must include a substratum common to it with the old religions of the Turanians and Aryans. Neglect of these very simple considerations often leads to great confusion in the minds both of Christians and unbelievers, as to the relation of Christianity to heathenism, and especially to the older and more primitive forms of heathenism.

The Turanian stock, of which the Mongolian peoples of Northern Asia may be taken as the type, includes also the American races, and the oldest historical populations of Western Asia and of Europe ; and they are the peoples who, in their physical features and their art tendencies, most nearly resemble the prehistoric men of the caves and gravels. They largely consist of the populations which the Bible affiliates with Ham. They are remarkable for their permanent and stationary forms of civilization or barbarism, and for the languages least developed in grammatical structure. These people had and still have traditions of the creation and early history of man similar to those in the earlier Biblical books ; but the connection of their religions with that of the Bible breaks off from the time of Abraham ; and the earlier portions of revelation which they possessed became disintegrated into a polytheism which takes very largely the form of animism, or of attributing some special spiritual indwelling to all natural objects, and also that of worship of ancestors and heroes. The portion of primitive theological belief to which they have clung most persistently is the doctrine of the immortality of the soul, which in all their religious beliefs occupies a prominent place, and has always been connected with special attention to rites of sepulture and monuments to the dead. Their version of the revelation of creation appears most distinctly in the sacred book of the Quichés of Central America, and in the creation myths of the Mexicans, Iroquois, Algonquins, and other North American tribes ; and it has been handed down to us through the Semitic Assyrians from the ancient Chaldæo-turanian population of the valley of the Euphrates.

The Aryan races have been remarkable for their changeable and versatile character. Their religious ideas in the most primitive times appear to have been not dissimilar from those of the Turanians ; and the Indians, Persians, Greeks, Scandinavians, and Celts have all gone some length in developing and modifying these, apparently by purely human imaginative and intellectual materials. But all these developments were defective in a moral point of view, and had lost the stability and rational basis which proceed from monotheism. Hence they have given way before other and higher faiths; and at this day the more advanced nations of the Aryan, or in Scriptural language the Japhetic stock, have adopted the Semitic faith; and, as Noah long ago predicted, "dwell in the tents of Shem." No indigenous account of the genesis of things remains among the Aryan races, with the exception of that in the Avesta, and in some ancient Hindoo hymns, and these are merely variations of the Turanian or Semitic cosmogony. God has given to the Aryans no special revelations of his will, and they would have been left to grope for themselves along the paths of science and philosophy, but for the advent among them of the prophets of "Jehovah the God of Shem."

It is to the Semitic race that God has been most liberal in his gift of inspiration. Gathering up and treasuring the old common inheritance of religion, and eliminating from it the accretions of superstition, the children of Abraham at one time stood alone, or almost alone, as adherents of a belief in one God the Creator.   Their theology was added to from age to age by a succession of prophets, all working in one line of development, till it culminated in the appearance of Jesus Christ, and then proceeded to expand itself over the other races.   Among them it has undergone two remarkable phases of retrograde development — the one in Mohammedanism, which carries it back to a resemblance to its own earlier patriarchal stage, the other in Roman and Greek ecclesiasticism, which have taken it back to the Levitical system, along with a strong color of paganism.   Still its original documents survive, and retain their hold on large portions of the more enlightened Aryan nations, while through their means these documents have entered on a new career of conquest among the Semites and Turanians.   They are, however, it must be admitted, among the Aryan races of Europe, growing in a somewhat uncongenial soil ; partly because of the materialistic organization of these races, and partly because of the abundant remains of heathenism which still linger among them ; and it is possible that they may not realize their full triumphs over humanity till the Semitic races return to the position of Abraham, and erect again in the world the standard of monotheistic faith, under the auspices of a purified Christianity.

It follows from this hasty survey that it is the Semitic solution of the question of origins, as contained in the Hebrew Scriptures, that mainly concerns us ; and in the first place we must consider the foundation and historical development of this solution, as many misconceptions prevail on these points.   We may discuss these subjects under the heads of the Abrahamic Genesis and the Mosaic Genesis, and may in a subsequent chapter consider the results of these in the Genesis of the later Scripture writers.

THE ABRAHAMIC GENESIS.

It has been a favorite theory with some learned men that the earlier parts of the book of Genesis existed as ancient documents even in the time of Moses, and were incorporated by him in his work, and attempts have been made to separate, on various grounds, the older from the newer portions. Until lately, however, these attempts have been altogether conjectural and destitute of any positive basis of archæological fact. A new and interesting aspect has been given to them by the recent readings of the inscriptions on clay tablets found at Nineveh, and to which especial attention has been given by the late Mr. G. Smith, of the Archæological Department of the British Museum.

Assurbanipal, king of Assyria, one of the kings known to the Greeks by the name of Sardanapalus, reigned at Nineveh about B.C. 673. He was a grandson of the Biblical Sennacherib, and son of Esarhaddon, and it seems that he had inherited from his fathers a library of Chaldean and Assyrian literature, written not on perishable paper or parchment, but on tablets of clay, and containing much ancient lore of the nations inhabiting the valleys of the Tigris and Euphrates. Assurbanipal, living when the Assyrian empire had attained to the acme of its greatness, had leisure to become a greater patron of learning than any preceding king. His scribes ransacked the record chambers of the oldest temples in the world ; and Babel, Erech, Accad, and Ur had to yield up their treasures of history and theology to diligent copyists, who transcribed them in beautiful arrow-head characters on new clay tablets, and deposited them in the library of the great king.. It would appear that, at the same time,, these documents were edited, archaic forms of expression translated, and lacunæ caused by decay or fracture repaired. They were also inscribed with legends stating the sources whence they had been derived.

The empire of Assyria went down in blood, and its palaces were destroyed with fire, but the imperishable clay tablets which had formed the treasure of their libraries remained, more or less broken it is true, among the ruins. Exhumed by Layard and Smith, they are now among the collections of the British Museum, and their decipherment is throwing a new and strange light on the cosmogony and religions of the early East. Though the date of the writing of these tablets is comparatively modern, being about the time of the later kings of Judah, the original records from which they were transcribed profess to have been very ancient—some of them about 1600 years before the time of Assurbanipal, so that they go back to a time anterior to that of the early Hebrew patriarchs. Their genuineness has been endorsed, in one case, by the discovery by Mr. Loftus, in the city of Senkereh, of an apparent original, bearing date about 1600 years before Christ, and other inscriptions of equal or greater antiquity have been found in the ruins of Ur, on the Euphrates. Nor does there seem any reason to doubt that the scribes of Assurbanipal faithfully transcribed the oldest records extant in their time. Their care and diligence are also shown by the fact that where different versions of these records existed in different cities, they have made copies of these variant manuscripts, instead of attempting to reduce them to one text. The subjects treated of in the Nineveh tablets are very various, but those that concern our present purpose are the documents relating to the creation, the fall of man, and the deluge, of which considerable portions have been recovered, and have been translated by Mr. Smith.

These documents carry us back to a time when the Turanian religions had not yet been separated from the Semitic. The early Chaldeans, termed Cushites in the Bible, and who under Nimrod seem to have established the first empire in that region, are now known to have been Turanian; and among them apparently arose at a very early period a literature and a mythology. The Chaldeans were politically subjugated by the Semitic Assyrians, but they retained their religious predominance; and until a comparatively late period existed as a learned and priestly caste. To these primitive *Chasdim* were undoubtedly due the creation legends collected by the scribes of Assurbanipal. They were obtained in the old Chaldean cities, in the temples under the guardianship of Chaldean priests; and their date carries them back to a time anterior to the Assyrian conquest, and in which Chaldean kings still reigned. Here, then, we have an important connecting link between the cosmogonies of the Turanian and Semitic races; and leaving out of sight for the present the legends of the deluge and other matters allied to it, we may inquire as to the nature and contents of the Assyrian and Chaldean record of creation.

The Assyrian Genesis is similar in order and arrangement to that in our own Bible, and gives the same general order of the creative work. Its days, however, of creation, as indeed there is good internal evidence to prove those of Moses also are, seem to be periods or ages. It treats of the creation of gods, as well as of the universe, and thus introduces a polytheistic system; and it seems to recognize, like the Avesta, a primitive principle of evil, presiding over chaos, and subsequently introducing evil among men. These points may be illustrated by an extract from Mr. Smith's translation.   It relates to the earlier part of the work :

" When above were not raised the heavens,
  And below on the earth a plant had not grown up
  The deep also had not broken up its boundaries
  Chaos (or water) Tiamat (the sea or abyss) was the producing mother
     of them all
  These waters at the beginning were ordained
  But a tree had not grown a flower had not unfolded
  When the gods had not sprung up any one of them
  A plant had not grown and order did not exist
  Were made also the great gods
  The gods Lahma and Lahamu they caused to come * * *
  And they grew * * *
  The gods Sar and Kisar were made
  A course of days and a long time passed
  The god Anu * * *
  The gods Sar and * * * "

Here the first existences are Chaos (Mummu, or confusion) and Tiamat, which is the Thalatth of Berosus, representing the sea or primitive abyss, but also recognized as a female deity or first mother.   Then we have Lahma and Lahamu, which represent power or motion in nature, and are the equivalents of the Divine Spirit moving on the face of the waters in our Genesis.   Next we have the production of Sar or Iloar and Kisar, representing the expanse or firmament.   Sar is supposed to be the god Assur of the Assyrians, a great weather god, and after whom their nation and its founder were named.   The next process is the creation of the heaven and the earth, represented by Anu and Anatu.   Anu was always one of the greater gods, and was identified with the higher or starry heavens.   In succeeding tablets to this we find Bel or Belus introduced, as the agent in the creation of

animals and of men ; and he is the true Demiurgus or Mediator of the Assyrian system.   Next we have the introduction
of Hea or Saturn, who is the equivalent of the Biblical Adam,
and of Ishtar, mother of men, who is the Isha or Eve of
Genesis.   The rest of this legend evidently relates to deified men, among whom are Merodach, Nebo, and other heroes.

The first remark that we may make on this Assyrian Genesis is that, while it resembles generally the Mosaic account of
creation, it also strongly resembles the old cosmogonies of
the Egyptians and Persians, and those of the widely scattered
Turanians of Northern Asia and of America.   As an extreme
illustration of this, and to obviate the necessity of digression
at this point of our inquiry, I introduce here some extracts
from the Popul Vuh, or sacred book of the Quiché Indians
of Central America, an undoubted product of prehistoric religion in the western continent.*

"And the heaven was formed, and all the signs thereof set in their
angle and alignment, and its boundaries fixed toward the four winds by
the Creator and Former, and Mother and Father of life and existence—he
by whom all move and breathe, the Father and Cherisher of the peace of
nations and of the civilization of his people—he whose wisdom has pro-
jected the excellence of all that is on the earth or in the lakes or in the
sea."

"Behold the first word and the first discourse.   There was yet no man
nor any animal, * * * nothing was but the firmament.   The face of the earth
had not yet appeared over the peaceful sea, and all the space of heaven
* * * nothing but immobility and silence in the night."

"Alone also the Creator, the Former, the Dominator, the Feathered
Serpent—those that engender, those that give being—they are upon the

* I avail myself of the condensed translation in Bancroft's "Native
Races," vol. iii.   The original French translation of Brasseur du Bourbourg is more full.

water like a growing light. They are enveloped in green and blue, and therefore their name is Gucumatz."*

"Lo now how the heavens exist, how exists also the Heart of Heaven; such is the name of God. It is thus that he is called. And they spake, they consulted together and meditated; they mingled their words and their opinions."

"And the creation [of the earth] was verily after this wise. Earth, they said, and on the instant it was formed; like a cloud or a fog was its beginning. Then the mountains rose over the water like great fishes; in an instant the mountains and the plains were visible, and the cypress and the pine appeared. Then was the Gucumatz filled with joy, crying out: Blessed be thy coming, O Heart of Heaven, Hurakan, Thunderbolt. Our work and our labor has accomplished its end."

This corresponds to the work of the first four creative days; and next details are given as to the introduction of animals, with which, however, the Creator is represented as dissatisfied, because they could not know or invoke the Creator. They are therefore condemned to be subject to be devoured one of another. Again there is a council in heaven, and the gods determine to make man. But he also is imperfect, for he has speech without intelligence: so he is condemned to be destroyed by water. A new council is held, and a second race of men produced; but this fails in the capacity for religious worship—"they forgot the Heart of Heaven." These were partly destroyed by fire and partly converted into apes. Lastly another council is held, and perfect men created. Then follows a remarkable series of stories relating to the early history and migrations of men.

It is known that similar creation myths existed among the

---

* The Feathered Serpent is perhaps the representative of the Dragon and Serpent in the Semitic version; but has not the same evil import, and his color gave sacredness to blue and green stones, as the turquois and emerald, both in North and South America, and perhaps also in Asia and Africa.

Mexicans and other early civilized nations of America, and
in ruder and more grotesque forms even among the semi-barbarous and hunter tribes.   Their connection with the ancient
Semitic and Turanian revelations of Asia is unquestionable.

We have thus in the Assyrian Genesis a relic of early religious belief belonging to a period when such widely separated
stocks as the Assyrian and American were still one : to a
period, therefore, presumably long anterior to that of Moses.
Yet at this very early period the central portions at least of
the Turanian race had already devised some means of record-
ing their traditions in writing—probably the arrow-head writing, afterwards used by the Assyrians, had already been invented.   Again, at this early period a complex polytheism
had already sprung up, and this was connected with cosmological ideas, inasmuch as the primitive abyss, the firmament,
the starry heavens, the principle of life, were all subordinate
gods ; and so were also some of the earliest of the patriarchs
of the human race.   It is possible, however, that this was
among the early Chaldeans an exoteric representation for the
vulgar, and that the priestly caste may have understood it in
a monotheistic sense.   In any case, the idea of a Supreme
Creator remains behind the whole.   Farther, in the early
Chaldean record we have a more detailed and expanded
document than that of the Hebrew Genesis, probably intended for the popular ear, and to include as much as possible of
the current mythology.   As an example, I quote the following in relation to the creation of the moon, being apparently
a part of the narrative of that creative period corresponding
with the fourth day of Genesis :

"In its mass [that is, of the lower chaos] he made a boiling,
  The God Uru [the moon] he caused to rise out, the night he over-
    shadowed.

To fix it also for the light of the night until the shining of the day,
That the month might not be broken and in its amount be regular.
At the beginning of the month at the rising of the night,
His horns are breaking through to shine in the heavens.
On the seventh day to a circle he begins to swell,
And stretches toward the dawn farther."

We now come to the historical connection of all this with Abraham and with the Hebrew Scriptures. The early life of the "Father of the Faithful" belongs to the time when Turanian and Semitic elements were mingled in the Euphratean valley. Himself of the stock of Shem, he dwelt in Ur of the Chaldees, a city in whose ruins, now known by the name of Mugheir, Chaldean inscriptions have been found of a date anterior to that of the patriarch. In the time of Abraham a polytheistic religion already existed in Ur, for we are told that his father "served other gods." Further, the legends of the creation and the deluge, and the antediluvian age, with the history of Nimrod and other postdiluvian heroes, existed in a written form; and, strange though this may seem, there can be little doubt that Abraham, before he left Ur of the Chaldees, had read the same creation legends that have so recently been translated and published by Mr. Smith. But Abraham's relation to these was of a peculiar kind. With a spiritual enlightenment beyond that of his age, he dissented from the Turanian animism and polytheism, and maintained that pure and spiritual monotheism which, according to the Bible, had been the original faith of the sons of Noah. But he was overborne by the tendencies of his time, and probably by the royal and priestly influence then dominant in Chaldea, and he went forth from his native land in search of a country where he might have freedom to worship God. It is thus that Abraham appears as the earliest reformer, the first of

B

those martyrs of conscience who fear not to differ from the majority, the father and prototype of the faithful of every age, and the earliest apostle of the monotheistic faith which still reigns among all the higher races of men.

Did Abraham take with him in his pilgrimage the records of his people? It is scarcely possible to doubt that he did, and this probably in a written form, but purified from the polytheism and inane imaginations accreted upon them ; or perhaps he had access to still older and more primitive records anterior to the rise of the Turanian superstitions. In any case we may safely infer that Abraham and his tribe carried with them the substance of all that part of Genesis which contains the history of the world up to his time, and that this would be a precious heir-loom of his family, until it was edited and incorporated in the Pentateuch by his great descendant Moses. It seems plain, therefore, that the original prophet or seer to whom the narrative of creation was revealed lived before Abraham, but we need not doubt that the latter had the benefit of divine guidance in his noble stand against the idolatry of his age, and in his selection of the documents on which his own theology was based. These considerations help us to understand the persistence of Hebrew monotheism in the presence of the idolatries of Canaan and Egypt, since these were closely allied to the Chaldean system against which Abraham had protested. They also explain the recognition by Abraham, as co-religionists, of such monotheistic personages as Melchisedec, king of Salem. They further illustrate the nature of the religious basis in his people's beliefs on which Moses had to work, and on which he founded his theocratic system.

Before leaving this part of the subject, I would observe that the view above given, while it explains the agreement be-

tween the Hebrew Genesis and other ancient religious beliefs, is in strict accordance with the teachings of Genesis itself. The history given there implies monotheism and knowledge of God as the Creator and Redeemer, in antediluvian
and early postdiluvian times, a decadence from this into a systematic polytheism at a very early date, the protest and dissent of Abraham, his call of God to be the upholder of a purer
faith, and the maintenance of that faith by his descendants.
Besides this, any careful reader of Genesis and of the book of
Job, which, whatever its origin, must be more ancient than the
Mosaic law, will readily discover indications that Abraham
and the patriarchs were in the possession of documents and
traditions of the same purport with those in the early chapters
of Genesis, and that these were to them their only sacred literature. The reader of the Pentateuch must carry this idea
with him, if he would have any clear conception of the unity
and symmetry of these remarkable books.

THE MOSAIC GENESIS.

In the period of 400 years intervening between Abraham's
departure from Ur and the exodus of Israel from Egypt, no
great prophetic mind, like that of the Father of the Faithful,
appeared among the Hebrews. But then arose Moses, the
greatest figure in all antiquity before the advent of Christ,
and who was destined to give permanence and world-wide
prevalence to the faith for which Abraham had sacrificed so
much. Under the leadership of Moses, the Abrahamidæ,
now reduced to the condition of a serf population, emancipated themselves from Egyptian bondage, and, after forty
years of wandering desert life, settled themselves permanently on the hills and in the valleys of Palestine. The voice of
the ruling race, indistinctly conveyed to us from that distant

antiquity, maintains that the fugitive slaves were an abject and
contemptible herd ; but the leader of the exodus informs us
that, though cruelly trodden down by a haughty despot, they
were of noble parentage, the heirs of high hopes and promises.   Their migration is certainly the most remarkable national movement in the world's history—remarkable, not merely
in its events and immediate circumstances, but in its remote
political, literary, and moral results.   The rulers of Egypt,
polished, enlightened, and practical men, were yet the devo-
tees of a complicated system of hero and animal worship, like
that from which Abraham dissented, and derived in great part
from the " animism " which caused some of the oldest nations
of the world to associate a spiritual indwelling with the natural objects surrounding them ; or, if they had ceased to believe in this, they had sunk into a materialistic devotion to
the good things of the present world, combined with a superstitious belief in the efficacy of priestly absolution.

The slaves, leaving all this behind them, rose in their religious opinions to the pure and spiritual monotheism of the
great father of their race ; and their leader presented to them
a law unequalled up to our time in its union of justice, patriotism, and benevolence, and established among them, for the
first time in the world's history, a free constitutional republic.
Nor is this all ; unexampled though such results are elsewhere
in the case of serfs suddenly emancipated.   The Hebrew lawgiver has interwoven his institutions in a great historical
composition, including the grand and simple cosmogony of
the patriarchs, a detailed account of the affiliation and ethnological relations of the races of men, and a narrative of the
fortunes of his own people ; intimating not only that they
were a favored and chosen race, but that of them was to arise
a great Deliverer, who would bless all nations with pardon

and with peace,* and would solve once for all those great problems of the relations of man to God and the unseen world, which in the time of Moses as in our own were the most momentous of all, and gave to questions of origins all their practical value.

The lawgiver passed to his rest. His laws and literature, surviving through many vicissitudes, have produced in each succeeding age a new harvest of poetry and history, leavened with their own spirit. In the mean time the learning and the superstition of Egypt faded from the eyes of men. The splendid political and military organizations of Assyria, Babylon, Persia, and Macedon arose and crumbled into dust. The wonderful literature of Greece blazed forth and expired. That of Rome, a reflex and copy of the former, had reached its culminating point ; and no prophet had arisen among any of these Gentile nations to teach them the truth of God. The world, with all its national liberties crushed out, its religion and its philosophy corrupted and enfeebled to the last degree by an endless succession of borrowings and intermixtures, lay prostrate under the iron heel of Rome. Then appeared among the now obscure remnant of Israel, one who announced himself as the Prophet like unto Moses, promised of old ; but a prophet whose mission it was to redeem not Israel only, but the whole world, and to make all who will believe, children of faithful Abraham. Adopting the whole of the sacred literature of the Hebrews, and proving his mission by its words, he sent forth a few plain men to

---

* I do not think it necessary to attach any value to the doubts of certain schools of criticism as to the Mosaic authorship of the Pentateuch. Whatever quibbles may be raised on isolated texts, no rational student can doubt that we have in these books a collection of authentic documents of the Exodus. They are absolutely inexplicable on any other supposition.

write its closing books, and to plant it on the ruins of all the time-honored beliefs of the nations—beliefs supported by a splendid and highly organized priestly system and by despotic power, and gilded by all the highest efforts of poetry and art.

The story is a very familiar one ; but it is marvellous beyond all others. Nor is the modern history of the Bible less wonderful. Exhumed from the rubbish of the Middle Ages, it has entered on a new career of victory. It has stimulated the mind of modern Europe to all its highest efforts, and has been the charter of its civil and religious liberties. Its wondrous revelation of all that man most desires to know, in the past, in the present, and in his future destinies, has gone home to the hearts of men in all ranks of society and in all countries. In many great nations it is the only rule of religious faith. In every civilized country it is the basis of all that is most valuable in religion. Where it has been withheld from the people, civilization in its highest aspects has languished, and superstition, priestcraft, and tyranny have held their ground or have perished under the assaults of a heartless and inhuman infidelity. Where it has been a household book, education has necessarily flourished, liberty has taken root, and the higher nature of man has been developed to the full. Driven from many other countries by tyrannical interference with liberty of thought and discussion, or by a shortsighted ecclesiasticism, it has taken up its special abode with the greatest commercial nations of our time ; and, scattered by their agency broadcast over the world, it is read by every nation under heaven in its own tongue, and is slowly but surely preparing the way for wider and greater changes than any that have heretofore resulted from its influence. Explain it as we may, the Bible is a great literary miracle ; and no amount of inspiration or authority that can be claimed for it is more strange or incredible than the actual history of the book. Yet no book has ever thrown itself into so decided antagonism with all the great forces of evil in the world. Tyranny hates it, because the Bible so strongly maintains the individual value and rights of man as man. The spirit of caste dislikes it for the same reason. Anarchical license, on the other hand, finds nothing but discouragement in it. Priestcraft gnashes its teeth at it, as the very embodiment of private judgment in religion, and because it so scornfully ignores human authority in matters of conscience, and human intervention between man and his Maker. Skepticism sneers at it, because it requires faith and humility, and threatens ruin to the unbeliever. It launches its thunders against every form of violence or fraud or allurement that seeks to profit by wrong or to pander to the vices of mankind ; all these consequently are its foes. On the other hand, by its uncompromising stand with reference to certain scientific and historical facts, it has appeared to oppose the progress of thought and speculation ;.though, as we shall see, it has been unfairly accused in this last respect.

With its antagonism to the evil that is in the world we have at present nothing to do, except to caution the student of this venerable literature against the prejudices which interested and unscrupulous foes seek to cultivate. Its doctrine of the origin of man and of the world, and the relation of this to modern scientific and historical results, is that which now claims our attention ; and this more especially in the relation which the Mosaic cosmogony, considered as an early revelation from God, may be found to bear to the facts which modern scientific research has elicited from the universe itself. The aspects in which apparent conflicts present

themselves are threefold.   At one time it was not unusual to
impugn the historical accuracy of the Pentateuch on the
evidence of the Greek historians ; and on many points
scarcely any corroborative evidence could be cited in favor
of the Hebrew writers.   In our own time much of this difficulty has been removed, and an immense amount of learned
research has been reduced to waste paper, by the circumstance that the monuments of Egypt and Assyria have risen
up to bear testimony in favor of the Bible ; and scarcely any
sane man now doubts the value of the Hebrew history.   The
battle-ground has in consequence been shifted farther back,
to points concerning the affiliation of the races of men, the
absolute antiquity of man's residence on the earth, and the
condition of prehistoric men ; questions on which we can
scarcely expect to find, at least for a long time, any decisive monumental or scientific evidence.   Secondly, the
Bible commits itself to certain cosmological doctrines and
statements respecting the system of nature, and details of
that system, more or less approaching to the domain which
geology occupies in its investigations of the past history of
the earth ; and at every stage in the progress of modern
science, independently of the mischief done by smatterers and
skeptics, earnest bigotry on the one hand, and earnest scientific enthusiasm on the other, have come into collision.   One
stumbling-block after another has, it is true, been removed
by mutual concession and farther enlightenment, and by the
removal of false traditional interpretations of the sacred records, as well as by farther discoveries in relation to nature.
But the field of conflict has thereby apparently only changed ;
and we still have some Christians in consequence regarding
the revelations of natural science with suspicion, and some
scientific men cherishing a sullen resentment against what

they regard as an intolerant intermeddling of theology with the domain of legitimate investigation. Lastly, the great growth of physical science, and the tendency to take partial views of the universe as if it were comprehended in mere matter and force, with similarly partial views of the doctrines of continuity and the conservation of forces, along with the growth of a belief in spontaneous evolution as a philosophical dogma, have placed many scientific minds in a position which makes them treat the whole question of the origin and destiny of man and of the world with absolute indifference.

There can nevertheless be no question that the whole subject is at the present moment in a more satisfactory state than ever previously ; that much has been done for the solution of difficulties ; that many theologians admit the great service which in many cases science has rendered to the interpretation of the Bible, and that most naturalists feel themselves free from undue trammels. Above all, there is a very general disposition to admit the distinctness and independence of the fields of revelation and natural science, the possibility of their arriving at some of the same truths, though in very different ways, and the folly of expecting them fully and manifestly to agree in the present state of our information. The literature of this kind of natural history has also become very extensive, and there are few persons who do not at least know that there are methods of reconciling the cosmogony of Moses with that obtained from the study of nature. For this very reason the time is favorable for an unprejudiced discussion of the questions involved ; and for presenting on the one hand to naturalists a summary of what the Bible does actually teach respecting the early history of the earth and man, and on the other to those whose studies lie in the book which they regard as the Word of God, rather than in the

material universe which they regard as his work, a view of
the points in which the teaching of the Bible comes into
contact with natural science at its present stage of progress.
These are the ends which I propose to myself in the following pages, and which I shall endeavor to pursue in a spirit of
fair and truthful investigation; having regard on the one
hand to the claims and influence of the venerable Book of
God, and on the other to the rights and legitimate results of
modern scientific inquiry.

The plan which I have proposed to myself in this part of
my subject is to take the statements of Genesis in their order,
and consider what they import, and how they appear to harmonize with what we know from other sources. This will
occupy some space, but it will save time in dealing with the
remaining parts of the subject. Before entering upon it, I
propose to devote one chapter to the answers to three questions which concern the whole doctrine of revealed religion,
whether Semitic, Turanian, or Aryan. These are: (1) *Why*
the origin of things should be revealed; (2) *How* it could be
revealed; and (3) *What* would require to be revealed in order to form the basis of a rational theism.

CHAPTER II.

OBJECTS AND NATURE OF A REVELATION OF ORIGINS.

"There are two books from which I collect my divinity; besides that written one of God, another of his servant nature—that universal and public manuscript that lies expansed unto the eyes of all."—SIR T. BROWNE.

THERE are some questions, simple enough in themselves, respecting the general character and object of the references to nature and creation in the Scriptures, which yet are so variously and vaguely answered that they deserve some consideration before entering on the detailed study of the subject. These are : (1) The object of the introduction of such subjects into the Hebrew sacred books—the *why* of the revelation of origins. (2) The origin, character, and structure of the narrative of creation and other cosmological statements in those books—the *how* of the revelation. (3) The character of the Biblical cosmogony, and general views of nature to which it leads—the *what* of the revelation.

(1) *The Object of the Introduction of a Cosmogony in the Bible.* —Man, even in his rudest and most uncivilized state, does not limit his mental vision to his daily wants. He desires to live not merely in the present, but in the future also and the past. This is a psychological peculiarity which, as much as any other, marks his separation from the lower animals, and which in his utmost degradation he never wholly loses. Whatever may be fancied as to imagined prehistoric nations, it is certain that no people now existing, or historically known to us,

is so rude as to be destitute of some hopes or fears in reference to the future, some traditions as to the distant past.
Every religious system that has had any influence over the
human mind has included such ideas.   Nor are we to regard
this as an accident.   It depends on fixed principles in our
constitution, which crave as their proper aliment such information; and if it can not be obtained, the mind, rather
than want it, invents for itself.   We might infer from this
very circumstance that a true religion, emanating from the
Creator, would supply this craving; and might content ourselves with affirming that, on this ground alone, it behooved
revelation to have a cosmogony.

But the religion of the Hebrews especially required to be
explicit as to the origin of the earth and all things therein.
Its peculiar dogma is that of one only God, the Creator, requiring the sole homage of his creatures.   The heathen for
the most part acknowledged in some form a supreme god,
but they also gave divine honors to subordinate gods, to deceased ancestors and heroes, and to natural phenomena, in
such a manner as practically to obscure their ideas of the
Creator, or altogether to set aside his worship.   The influence of such idolatry was the chief antagonism which the
Hebrew monotheism had to encounter; and we learn from
the history of the nation how often the worshippers of Jehovah
were led astray by its allurements.   To guard against this
danger, it was absolutely necessary that no place should be
left for the introduction of polytheism, by placing the whole
work of creation and providence under the sole jurisdiction
of the One God.   Moses consequently takes strong ground
on these points.   He first insists on the creation of all things
by the fiat of the Supreme.   Next he specifies the elaboration
and arrangement of all the powers of inanimate nature, and

the introduction of every form of organic existence, as the work of the same First Cause. Lastly, he insists on the creation of a primal human pair, and on the descent from them of all the branches of the human race, including of course those ancestors and magnates who up to his time had been honored with apotheosis; and on the same principle he explains the golden age of Eden, the fall, the cherubic emblems, the deluge, and other facts in human history interwoven by the heathen with their idolatries. He thus grasps the whole material of ancient idolatry, reduces it within the compass of monotheism, and shows its relation to the one true primitive religion, which was that not only of the Hebrews, but of right that of the whole world, whose prevailing polytheism consisted in perversions of its truth or unity. For such reasons the early chapters of Genesis are so far from being of the character of digressions from the scope and intention of the book, that they form a substratum of doctrine absolutely essential to the Hebrew faith, and equally so to its development in Christianity.

The references to nature in the Bible, however, and especially in its poetical books, far exceed the absolute requirements of the reasons above stated ; and this leads to another and very interesting view, namely, the tendency of monotheism to the development of truthful and exalted ideas of nature. The Hebrew theology allowed no attempt at visible representations of the Creator or of his works for purposes of worship. It thus to a great extent prevented that connection of imitative art with religion which flourished in heathen antiquity, and has been introduced into certain forms of Christianity. But it cultivated the higher arts of poetry and song, and taught them to draw their inspiration from nature as the only visible revelation of Deity. Hence the growth of a healthy "physico-theology," excluding all idolatry of natural phenomena, and all superstitious dread of them as independent powers, but inviting to their examination as manifestations of God, and leading to conceptions of the unity of plan in the cosmos, of which polytheism, even in its highest literary efforts, was quite incapable. In the same manner the Bible has always proved itself an active stimulant of natural science, connecting such studies, as it does, with our higher religious sentiments; while polytheism and materialism have acted as repressive influences, the one because it obscures the unity of nature, the other because, in robbing it of its presiding Divinity, it gives a cold and repulsive, corpse-like aspect, chilling to the imagination, and incapable of attracting the general mind.

Naturalists should not forget their obligations to the Bible in this respect, and should on this very ground prefer its teachings to those of modern pantheism and positivism, and still more to those of mere priestly authority. Very few minds are content with simple materialism, and those who must have a God, if they do not recognize the Jehovah of the Hebrew Scriptures as the Creator and Supreme Ruler of the universe, are too likely to seek for him in the dimness of human authority and tradition, or of pantheistic philosophy; both of them more akin to ancient heathenism than to modern civilization, and in their ultimate tendencies, if not in their immediate consequences, quite as hostile to progress in science as to evangelical Christianity.

Every student of human nature is aware of the influence in favor of the appreciation of natural beauty and sublimity which the Bible impresses on those who are deeply imbued with its teaching; even where that same teaching has induced what may be regarded as a puritanical dislike of imitative art, at least in its religious aspects. On the other hand, naturalists can not refuse to acknowledge the surpassing majesty of the views of nature presented in the Bible. No one has expressed this better than Humboldt : " It is characteristic of the poetry of the Hebrews that, as a reflex of monotheism, it always embraces the universe in its unity, comprising both terrestrial life and the luminous realms of space ; it dwells but rarely on the individuality of phenomena, preferring the contemplation of great masses. The Hebrew poet does not depict nature as a self - dependent object, glorious in its individual beauty, but always as in relation or subjection to a higher spiritual power. Nature is to him a work of creation and order—the living expression of the omnipresence of the Divinity in the visible world." In reference to the 104th Psalm, which may be viewed as a poetical version of the narrative of creation in Genesis, the same great writer remarks : " We are astonished to find in a lyrical poem of such a limited compass, the whole universe—the heavens and the earth—sketched with a few bold touches. The calm and toilsome life of man, from the rising of the sun to the setting of the same, when his daily work is done, is here contrasted with the moving life of the elements of nature. This contrast and generalization in the conception of the mutual action of natural phenomena, and the retrospection of an omnipresent invisible Power, which can renew the earth or crumble it to dust, constitute a solemn and exalted rather than a gentle form of poetic creation."*

If we admit the source of inspiration claimed by the Hebrew poets, we shall not be surprised that they should thus write of nature. We shall only lament that so many pious

---

* " Cosmos," Otté's translation.

and learned interpreters of Scripture have been too little acquainted with nature to appreciate the natural history of the
Book of God, or adequately to illustrate it to those who depend on their teaching ; and that so many naturalists have
contented themselves with wondering at the large general
views of the Hebrew poets, without considering that they are
based on a revelation of the nature and order of the creative
work which supplied to the Hebrew mind the place of those
geological wonders which have astonished and enlarged the
minds of modern nations.   A modern divine, himself well
read in nature, truly says : " If men of piety were also men
of science, and if men of science were to read the Scriptures,
there would be more faith on the earth and also more philosophy."*   In a similar strain the patient botanist of the
marine algæ thus pleads for the joint claims of the Bible and
nature : " Unfortunately it happens that in the educational
course prescribed to our divines natural history has no place,
for which reason many are ignorant of the important bearings
which the book of nature has on the book of revelation.
They do not consider, apparently, that both are from God—
both are his faithful witnesses to mankind.   And if this be so,
is it reasonable to suppose that either, without the other, can
be fully understood ?   It is only necessary to glance at the
absurd commentaries in reference to natural objects which
are to be found in too many annotations of the Holy Scriptures to be convinced of the benefit which the clergy would
themselves derive from a more extended study of the works
of creation.   And to missionaries especially, a minute familiarity with natural objects must be a powerful assistance in
awakening the attention of the savage, who, after his manner,

---

* Hamilton, " Royal Preacher."

is a close observer, and likely to detect a fallacy in his teacher, should the latter attempt a practical illustration of his discourse without sufficient knowledge. These are not days in which persons who ought to be our guides in matters of doctrine can afford to be behind the rest of the world in knowledge ; nor can they safely sneer at the knowledge which puffeth up, until, like the apostle, they have sounded its depths and proved its shallowness."* It is truly much to be desired that divines and commentators, instead of trying to distort the representations of nature in the Bible into the supposed requirements of a barbarous age, or of setting aside modern discoveries as if they could have no connection with Scripture truth, would study natural objects and laws sufficiently to bring themselves in this respect to the level of the Hebrew writers. Such knowledge would be cheaply purchased even by the sacrifice of a part of their verbal and literary training. It is well that this point is now attracting the attention of the Christian world, and it is but just to admit that some of our more eminent religious writers have produced noble examples of accurate illustrations of Scripture derived from nature. In any case, the Bible itself can not be charged with any neglect of the claims of nature or with any narrow tendency to place material and spiritual things in.antagonism to one another.

Another reason why a revelation from God must deal with the origins of things, is that such revelation is, like creation, in its own nature progressive. It is given little by little to successive generations of men, and must proceed from the first rudiments of religious truth onward to its higher developments with the growth of humanity from age to age. Hence

* Harvey, "Nereis Boreali Americana."

the teachings in the early chapters of Genesis are of the simplest and most child-like character, and the first of these early teachings is necessarily that of God the Creator, just as our elementary catechisms for children have been wont to begin with the question, "Who made you?" In this way man is led in the most direct and simple way to the feet of the Universal Father, and a foundation is laid whereon further religious teaching adapted to the growth of the individual mind and to the growing complications of human society can be built. But again, alike in the earliest and simplest as in the more advanced states of the human mind, if spiritual things are to be taught, it must be through the medium of material things. We have no language to express in any direct way spiritual truths; they must be given to us in terms of the natural. We have not yet learned the tongue of the immortals, and probably can not learn it in this world. The word "spirit" itself, which we borrow from the Latin, the Greek *Pneuma*, the Hebrew *Ruah*, primarily all agree in signifying breath or wind. We have to speak of our own breath when we mean our spiritual nature, of God's breath when we mean his spiritual nature, and so of all other things not obvious to our senses. There is constant danger in this that the material shall be taken for the spiritual of which it is the symbol, the figure for the reality, the creature for the Creator, and this danger is best counteracted by a decided testimony in relation to the origin of all material things in the will of the spiritual and eternal God. Thus the Bible writers are enabled to use a free and bold manner of speech respecting divine things. Their expressions at one time appear pantheistic and at another anthropomorphic; they see God in every thing, and use with the utmost freedom natural emblems to indicate his perfections and procedure, and our relations to him. In this way there is life and action in their teaching, and it is removed as far as possible from a dry, abstract theology, while equally remote from any tinge of idolatry or superstition.

It may, however, be objected that by the introduction of a cosmogony the Bible exposes itself to a conflict with science, and that thereby injury results both to science and to religion. This is a grave charge, and one that has evidently had much weight with many minds, since it has been the subject of entire treatises designed to illustrate the history of the conflict or to explain its nature. The revelation of God's will to man for his moral guidance, if necessary at all, was necessary before the rise of natural science. Men could not do without the knowledge of the unity of nature and of the unity of God, until these great truths could be worked out by scientific induction. Perhaps they might never have been so worked out. Therefore a revealed book of origins has a right to precedence in this matter. Nor need it in any way come into conflict with the science subsequently to grow up. Science does not deal so much with the origin of nature as with its method and laws, and all that is necessary on the part of a revelation, to avoid conflict with it, is to confine itself to statements of phenomena and to avoid hypotheses. This is eminently the course of the Bible. In its cosmogony it shuns all embellishments and details, and contents itself with the fact of creation and a slight sketch of its order; and in their subsequent references to nature the sacred writers are strictly phenomenal in their statements, and refer every thing directly to the will of God, without any theory as to secondary causes and relations. They are thus decided and positive on the points with reference to which it behooves revelation to testify, and absolutely non-committal on the points which belong to the exclusive domain of science.

What, then, are we to say of the imaginary " conflict of science with religion," of which so much has been made? Simply that it results largely from misapprehension and from misuse of terms. True religion, which consists in practical love to God and to our fellow-men, can have no conflict with science. True science is its fast ally. The Bible, considered as a revelation of spiritual truth to man for his salvation and enlightenment, can have no conflict with science. It promotes the study of nature, rendering it honorable by giving it the dignity of an inquiry into the ways of God, and rendering it safe by separating it from all ideas of magic and necromancy. It gives a theological basis to the ideas of the unity of nature and of natural law. The conflict of science, when historically analyzed, is found to have been fourfold—with the Church, with theology, with superstition, and with false or imperfect science and philosophy. Religious men may have identified themselves from time to time with these opponents, but that is all ; and much more frequently the opposition has been by bad men more or less professing religious objects. Organizations calling themselves " the Church," and whose warrant from the Bible is often of the slenderest, have denounced and opposed and persecuted new scientific truths ; but they have just as often denounced the Bible itself, and religious doctrines founded on it. Theology claims to be itself one of the sciences, and as such it is necessarily imperfect and progressive, and may at any time be more or less in conflict with other sciences ; but theology is not religion, and may often have very little in common either with true religion or the Bible. When discussions arise between theology and other sciences, it is only a pity that either side should indulge in what has been called the *odium theologicum*, but which is unfortunately not confined to divines. Superstition, considered as the unreasonable fear of natural agencies, is a passive rather than an active opponent of science. But revelation, which affirms unity, law, and a Father's hand in nature, is the deadly foe of superstition, and no people who have been readers of the Bible and imbued with its spirit have ever been found ready to molest or persecute science. Work of this sort has been done only by the ignorant, superstitious, and priest-ridden votaries of systems which withhold the Bible from the people, and detest it as much as they dislike science. Perhaps the most troublesome opposition to science, or rather to the progress of science, has sprung from the tenacity with which men hold to old ideas. These, which may have been at one time the best science attainable, root themselves in popular literature, and even in learned bodies and in educational books and institutions. They become identified with men's conceptions both of nature and religion, and modify their interpretations of the Bible itself. It thus becomes a most difficult matter to wrench them from men's minds, and their advocates are too apt to invoke in their defense political, social, and ecclesiastical powers, and to seek to support them by the authority of revelation, when this may perhaps be quite as favorable to the newer views opposed to them. All these conflicts are, however, necessary incidents in human progress, which comes only by conflict ; and there is reason to believe that they would be as severe in the absence of revealed religion as in its presence, were it not that the absence of revelation seems often to produce a fixity and stagnation of thought unfavorable to any new views, and consequently to some extent to any intellectual conflict. It has been, indeed, to the disinterment of the Bible in the Reformation of the fifteenth century that the world owes, more than to any other cause, the immense growth of modern science, and the freedom of discussion which now prevails. The Protestant idea of individual judgment in matters of religion is thoroughly Biblical, for the Bible everywhere appeals to men in this way; and this idea is the strongest guarantee that the world possesses for intellectual liberty in other matters.

We conclude, therefore, on all these grounds, that it was necessary that a revelation from God should take strong and positive ground on the question of the origin of the universe.

(2) *The Origin, Method, and Structure of the Scriptural Cosmogony.* — A respectable physicist, but somewhat shallow naturalist and theologian, whose works at one time attracted much attention, has said of the first chapter of Genesis : " It can not be history—it may be poetry." Its claims to be history we shall investigate under another head, but it is pertinent to our present inquiry to ask whether it can be poetry. That its substance or matter is poetical no one who has read it once can believe; but it can not be denied that in its form it approaches somewhat to that kind of thought-rhythm or parallelism which gives so peculiar a character to Hebrew poetry. We learn from many Scripture passages, especially in the Proverbs, that this poetical parallelism need not necessarily be connected with poetical thought; that in truth it might be used, as rhyme is sometimes with us, to aid the memory. The oldest acknowledged verse in Scripture is a case in point. Lamech, who lived before the flood, appears to have slain a man in self-defense, or at least in an encounter in which he himself was wounded; and he attempts to define the nature of the crime in the following words :

> "Adah and Zillah, hear my voice ;
> Ye wives of Lamech, hearken to my speech :—
> I have slain a man to my wounding,

> And a young man to my hurt;
> If Cain shall be avenged sevenfold,
> Truly Lamech seventy and seven fold."

All this is prosaic enough in matter, but the form into which it is thrown gives it a certain dignity, and impresses it on the memory; which last object was probably what the author of this sole fragment of antediluvian literature had in view. He succeeded too—for the sentiment was handed down, probably orally; and Moses incorporates it in his narration, perhaps on account of its interest as the first record of the distinction between willful murder like that of Cain, and justifiable homicide. It is interesting also to observe the same parallelism of style, no doubt with the same objects, in many old Egyptian monumental inscriptions, which, however grandiloquent, are scarcely poetical.* It also appears in that ancient record of creation and the deluge recently rescued from the clay tablets of Nineveh.

Now in the first chapter of Genesis, and the first three verses of chapter second, being the formal general narrative of creation, on which, as we shall see, every other statement on the subject in the Bible is based, we have this peculiar parallelism of style. If we ask why, the answer must, I think, be—to give dignity and symmetry to what would otherwise be a dry abstract, and still more to aid memory. This last consideration, perhaps indicating that this chapter, like the apology of Lamech, had been handed down orally for a long period, connects itself with the theory of the pre-Abrahamic origin of these documents to which reference has already been made.

The form of the narrative, however, in no way impairs its

---

* Osburn, "Monumental History of Egypt."

precision or accuracy of statement. On this Eichhorn well
says: "There lies at the foundation of the first chapter a
carefully designed plan, all whose parts are carried out with
much art, whereby its appropriate place is assigned to every
idea;" and we may add, whereby every idea is expressed in
the simplest and fewest words, yet with marvellous accuracy,
amounting to an almost scientific precision of diction, for
which both the form into which it is thrown and the homogeneous and simple character of the Hebrew language are
very well adapted. Much of this indeed remains in the English version, though our language is less perfectly suited
than the Hebrew for the concise announcement of general
truths of this description. Our translators have, however,
deviated greatly from the true sense of many important words,
especially where they have taken the Septuagint translation
for their guide, as in the words "firmament," "whales,"
"creeping things," etc. These errors will be noticed in subsequent pages. In the mean time I may merely add that the
labors of the ablest Biblical critics give us every reason to
conclude that the received text of Genesis preserves, almost
without an iota of change, the beautiful simplicity of its first
chapter; and that we now have it in a more perfect state
than that in which it was presented to the translators of most
of the early versions. It must also be admitted that the object in view was best served by that direct reference to the
creative fiat, and ignoring of all secondary causes, which are
conspicuous in this narrative. This is indeed the general
tone of the Bible in speaking of natural phenomena; and this
mode of proceeding is in perfect harmony with its claims to
divine authority. Had not this course been chosen, no other
could have been adopted, in strict consistency with truth,
short of a full revelation of the whole system of nature, in the

details of all its laws and processes. This we now know would have been impossible, and, if possible, useless or even mischievous.

Regarded from this point of view—the plenary inspiration of the book—the Scriptural references to creation profess to furnish a very general outline, for theological purposes, of the principal features of a vast region unexplored when they were written, and into which human research has yet penetrated along only a few lines. Natural science, in following out these lines of observation, has reached some of the objects delineated in the Scriptural sketch; of others it has obtained distant glimpses; many are probably unknown, and we can appreciate the true value and dimensions relatively to the whole of very few. So vast indeed are the subjects of the bold sketch of the Hebrew prophet, that natural science can not pretend as yet so to fill in the outline as quite to measure the accuracy of its proportions. Yet the lines, though few, are so boldly drawn, and with so much apparent unity and symmetry, that we almost involuntarily admit that they are accurate and complete. This may appear to be underrating the actual progress of science relatively to this great fore-shadowing outline; but I know that those most deeply versed in the knowledge of nature will be the least disposed to quarrel with it, whatever skepticism they may entertain as to the greater general completeness of the inspired record.

Another point which deserves a passing notice here is the theory of Dr. Kurtz and others, that the Mosaic narrative represents a vision of creation, analogous to those prophetic visions which appear in the later books of Scripture. This is beyond all question the most simple and probable solution of the origin of the document, when viewed as inspired, but we shall have to recur to it on a future page.

C

But with respect to the precise origin of this cosmogony, the question now arises, Is it really in substance a revelation from God to man? We must not disguise from ourselves that this deliberate statement of an order of creation in so far challenges comparison with the results of science, and this in a very different way from that which applies to the incidental references to nature in the Bible. Further, inasmuch as it relates to events which transpired before the creation of man, it is of the nature of prophecy rather than of history. It is, in short, either an inspired revelation of the divine procedure in creation, or it is a product of human imagination or research, or a deliberate fraud.

To no part of the Bible do these alternatives more strictly apply than to its first chapter. This "can not be history" in the strict acceptation of the term. It relates to events which no human eye witnessed, respecting which no human testimony could give any information. It represents the creation of man as the last of a long series of events, of which it professes to inform us. The knowledge of these events can not have been a matter of human experience. If at all entitled to confidence, the narrative must, therefore, be received as an inspired document, not handed down by any doubtful tradition, but existing as originally transfused into human language from the mind of the Author of nature himself. This view is in no way affected by the hypothesis, already mentioned, that the first chapters of Genesis were compiled by Moses from more ancient documents. This merely throws back the revelation to a higher antiquity, and requires us to suppose the agency of two inspired men instead of one.

It would be out of place here to enter into any argument for the inspiration of Scripture, or to attempt to define the nature of that inspiration. I merely wish to impress on the mind of the reader that without the admission of its reality, or at least its possibility, our present inquiry becomes merely a matter of curious antiquarian research. We must also on this ground distinguish between the claims of the Scriptures and those of tradition or secular history, when they refer to the same facts. The traditions and cosmogonies of some ancient nations have many features in common with the Bible narrative; and, on the supposition that Moses compiled from older documents, they may be portions of this more ancient sacred truth, but clothed in the varied garments of the fanciful mythological creeds which have sprung up in later and more degenerate times. Such fragments may safely be received as secondary aids to the understanding of the authentic record, but it would be folly to seek in them for the whole truth. They are but the scattered masses of ore, by tracing which we may sometimes open up new and rich portions of the vein of primitive lore from which they have been derived. It is, however, quite necessary here formally to inquire if there are any hypotheses short of that of plenary inspiration which may allow us to attach any value whatever to this most ancient document. I know but two views of this kind that are worthy of any attention.

1. The Mosaic account of creation may be a result of ancient scientific inquiries, analogous to those of modern geology.

2. It may be an allegorical or poetical mythus, not intended to be historical, but either devised for some extraneous purpose, or consisting of the conjectures of some gifted intellect.

These alternatives we may shortly consider, though the materials for their full discussion can be furnished only by facts to be subsequently stated. I am not aware that the first of these views has been maintained by any modern writer. Some eminent scientific men are, however, disposed to adopt such an explanation of the ancient Hindoo hymns, as well as of the cosmogony of Pythagoras, which bears evidence of this origin; and it may be an easy step to infer that the Hebrew cosmogony was derived from some similar source. Not many years ago such a supposition would have been regarded as almost insane. Then the science of antiquity was only another name for the philosophy of Greece and Rome. But in recent times we have seen Egypt disclose the ruins of a mighty civilization, more grand and massive though less elegant than that of Greece, and which had reached its acme ere Greece had received its alphabet—a civilization which, according to the Scripture history, is derived from that of the primeval Cushite empire, which extended from the plains of Shinar over all Southeastern Asia, but was crushed at its centre before the dawn of secular history. We have now little reason to doubt that Moses, when he studied the learning of Egypt, held converse with men who saw more clearly and deeply into nature's mysteries than did Thales or Pythagoras, or even Aristotle.* Still later the remnants of old

* On this subject I may refer naturalists to the intimate acquaintance with animals and their habits, indicated by the manner of their use as sacred emblems, and as symbols in hieroglyphic writing. Another illustration is afforded by the Mosaic narrative of the miracles and plagues connected with the exodus. The Egyptian king, on this occasion, consulted the *philosophers* and *augurs*. These learned men evidently regarded the serpent-rod miracle as but a more skilful form of one of the tricks of serpent-charmers. They showed Pharaoh the possibility of reddening the Nile water by artificial means, or perhaps by the development of red algæ in it. They explained the inroad of frogs on natural principles, probably referring to the immense abundance ordinarily of the ova and tadpoles of these creatures compared with that of the adults. But when the dust of the land became gnats ("lice" in our version), this was a phenomenon beyond their experience. Either the species was unknown to them, or its

Nineveh have been exhumed from their long sepulture, and antiquaries have been astonished by the discovery that knowledge and arts, supposed to belong exclusively to far more recent times, were in the days of the early Hebrew kings, and probably very long previously, firmly established on the banks of the Tigris. Such discoveries, when compared with hints furnished by the Scriptures, tend greatly to exalt our ideas of the state of civilization at the time when they were written ; and we shall perceive, in the course of our inquiry, many additional reasons for believing that the ancient Israelites were much farther advanced in natural science than is commonly supposed.

We have, however, no positive proof of such a theory, and it is subject to many grave objections. The narrative itself makes no pretension to a scientific origin, it quotes no authority, and it is connected with no philosophical speculations or deductions. It bears no internal evidence of having been the result of inductive inquiry, but appeals at once to faith in the truth of the great ultimate doctrine of absolute creation, and then proceeds to detail the steps of the process, in the manner of history as recorded by a witness, and not in the manner of science tracing back effects to their causes. Farther, it refers to conditions of our planet respecting which science has even now attained to no conclusions supported by evidence, and is not in a position to make dogmatic assertions. The tone of all the ancient cosmogonies has in

---

production out of the dry ground was an anomaly, or they knew that no larvæ adequate to explain it had previously existed. In the case of this plague, therefore, comparatively insignificant and easily simulated, they honestly confessed—"This is the finger of God." No better evidence could be desired that the savans here opposed to Moses were men of high character and extensive observation. Many other facts of similar tendency might be cited both from Moses and the Egyptian monuments.

these respects a resemblance to that of the Scriptures, and
bears testimony to a general impression pervading the mind
of antiquity that there was a divine and authoritative testimony to the facts of creation, distinct from history, philosophical speculation, or induction.

One of the boldest and simplest methods of this kind is
that followed by the authors of the "Types of Mankind," in
the attempt to assign a purely human origin to Genesis 1st.
These writers admit the greater antiquity of the first chapter,
though assigning the whole of the book to a comparatively
modern date.   They say:

"The 'document Jehovah'* does not especially concern
our present subject; and it is incomparable with the grander
conception of the more ancient and unknown writer of Genesis 1st.   With extreme felicity of diction and conciseness of
plan, the latter has defined the most philosophical views of
antiquity upon *cosmogony;* in fact so well that it has required
the palæontological discoveries of the nineteenth century—at
least 2500 years after his death—to overthrow his *septenary*
arrangement of 'Creation;' which, after all, would still be
correct enough in great principles, were it not for one individual oversight and one unlucky blunder; not exposed, however, until long after his era, by post-Copernican astronomy.
The oversight is where he wrote (Gen. i. 6–8), 'Let there be
raquiê,' *i. e.,* a *firmament;* which proves that his notions of
'sky' (solid like the concavity of a copper basin, with *stars*
set as brilliants in the metal) were the same as those of adjacent people of his time—indeed, of all men before the publication of Newton's 'Principia' and of Laplace's 'Mécanique
Céleste.'   The blunder is where he conceives that *aur,* 'light,'

---

* That in Genesis, chap. ii.

and *iom*, 'day' (Gen. i. 14–18), could have been physically possible *three whole days* before the 'two great luminaries,' *Sun* and *Moon*, were created. These venial errors deducted, his majestic song beautifully illustrates the simple process of ratiocination through which—often without the slightest historical proof of intercourse—different ''Types of Mankind,' at distinct epochas, and in countries widely apart, had arrived, naturally, at cosmogonic conclusions similar to the doctrines of that Hebraical school of which his harmonic and melodious numbers remain a magnificent memento.

"That process seems to have been the following : The ancients knew, as we do, that man *is* upon the earth ; and they were persuaded, as we are, that his appearance was preceded by unfathomable depths of time. Unable (as we are still) to measure periods antecedent to man by any *chronological* standard, the ancients rationally reached the tabulation of some events anterior to man through *induction*—a method not original with Lord Bacon, because known to St. Paul ; 'for his unseen things from the creation of the world, his power and Godhead, are clearly seen, *being understood by the things that are made*' (Rom. i., 20). Man, they felt, could not have lived upon earth without *animal* food ; ergo, 'cattle' preceded him, together with birds, reptiles, fishes, etc. Nothing living, they knew, could have existed without light and heat ; ergo, the *solar system* antedated animal life, no less than the *vegetation* indispensable for animal support. But terrestrial plants can not grow without *earth ;* ergo, that dry land had to be separated from pre-existent 'waters.' Their geological speculations inclining rather to the *Neptunian* than to the *Plutonian* theory—for Werner ever preceded Hutton— the ancients found it difficult to 'divide the waters from the waters' without interposing a metallic substance that 'divided

the waters which were *under* the firmament from the waters
that were *above* the firmament;' so they inferred, logically,
that a *firmament* must have been actually created for this object. [*E. g.*, ' The *windows* of the skies ' (Gen. vii., 11) ; 'the
waters *above* the skies ' (Psa. cxlviii., 4).] Before the 'waters '
(and here is the peculiar error of the genesiacal bard) some
of the ancients claimed the pre-existence of *light* (a view
adopted by the writer of Genesis 1st) ; while others asserted
that 'chaos' prevailed.   Both schools united, however, in the
conviction that DARKNESS—*Erebus*—anteceded all other *created things.*   What, said these ancients, can have existed before the 'darkness?'   *Ens entium*, the CREATOR, was the
humbled reply.   *Elohim* is the Hebrew vocal expression of
that climax ; to define whose attributes, save through the
phenomena of creation, is an attempt we leave to others more
presumptuous than ourselves."

The problem here set to the " unknown " author of Genesis is a hard one—given the one fact that " man is " to find
in detail how the world was formed in a series of preceding
ages of vast duration.   Is it possible that such a problem
could have been so worked out as to have endured the test
of three thousand years, and the scrutiny of modern science ?
But there is an "oversight" in one detail, and a "blunder"
in another.   By reference farther on, the reader will find under the chapters on "Light" and the "Atmosphere" that the
oversight and blunder are those not of the writer of Genesis,
but of the learned American ethnologists in the nineteenth
century ; a circumstance which cuts in two ways in defense
of the ancient author so unhappily unknown to his modern
critics.

The second of the alternatives above referred to, the mythical hypothesis, has been advanced and ably supported, es-

pecially on the continent of Europe, and by such English writers as are disposed to apply the methods of modern rationalistic criticism to the Bible. In one of its least objectionable forms it is thus stated by Professor Powell:

"The narrative, then, of six periods of creation, followed by a seventh similar period of rest and blessing, was clearly designed by adaptation to their conceptions to enforce upon the Israelites the institution of the Sabbath; and in whatever way its details may be interpreted, it can not be regarded as an *historical* statement of the *primeval* institution of a Sabbath; a supposition which is indeed on other grounds sufficiently improbable, though often adopted. * * * If, then, we would avoid the alternative of being compelled to admit what must amount to impugning the truth of those portions at least of the Old Testament, we surely are bound to give fair consideration to the only suggestion which can set us entirely free from all the difficulties arising from the geological contradiction which does and must exist against any conceivable interpretation which retains the assertion of the historical character of the details of the narrative, as referring to the distinct transactions of each of the seven periods. * * * The one great fact couched in the general assertion that all things were created by the sole power of one Supreme Being is the whole of the representation to which an historical character can be assigned. As to the particular form in which the descriptive narrative is conveyed, we merely affirm that it can not be history—it may be poetry."*

The general ground on which this view is entertained is the supposed irreconcilable contradiction between the literal interpretation of the Mosaic record and the facts of geology.

---

* Kitto's Cyclopædia, art. "Creation."

The real amount of this difficulty we are not, in the present
stage of our inquiry, prepared to estimate. We can, however,
readily understand that the hypothesis depends on the supposition that the narrative of creation is posterior in date to
the Mosaic ritual, and that this plain and circumstantial series
of statements is a fable designed to support the Sabbatical
institution, instead of the rite being, as represented in the
Bible itself, a commemoration of the previously recorded fact.
This is, fortunately, a gratuitous assumption, contrary to the
probable date of the documents, as deduced from internal
evidence and from comparison with the Assyrian and other
cosmogonies ; and it also completely ignores the other manifest uses mentioned under our first head. If proved, it would
give to the whole the character of a pious fraud, and would
obviously render any comparison with the geological history
of the earth altogether unnecessary. While, therefore, it
must be freely admitted that the Mosaic narrative can not
be history, in so far at least as history is a product of human
experience, we can not admit that it is a poetical mythus, or,
in other words, that it is destitute of substantial truth, unless
proved by good evidence to be so ; and, when this is proved,
we must also admit that it is quite undeserving of the credit
which it claims as a revelation from God.

Since, therefore, the events recorded in the first chapter of
Genesis were not witnessed by man; since there is no reason
to believe that they were discovered by scientific inquiry ;
and since, if true, they can not be a poetical myth, we must,
in the mean time, return to our former supposition that the
Mosaic cosmogony is a direct revelation from the Creator.
In this respect, the position of this part of the earth's Biblical
history resembles that of prophecy. Writers *may* accurately
relate contemporary events, or those which belong to the hu-

man period, without inspiration; but the moment that they profess accurately to foretell the history of the future, or to inform us of events which preceded the human period, we must either believe them to be inspired, or reject them as impostors or fanatics. Many attempts have been made to find intermediate standing-ground, but it is so precarious that the nicest of our modern critical balancers have been unable to maintain themselves upon it.

Having thus determined that the Mosaic cosmogony, in its grand general features, must either be inspired or worthless, we have further to inquire to what extent it is necessary to suppose that the particular details and mode of expression of the narrative, and the subsequent allusions to nature in the Bible, must be regarded as entitled to this position. We may conceive them to have been left to the discretion of the writers; and, in that case, they will merely represent the knowledge of nature actually existing at the time. On the other hand, their accuracy may have been secured by the divine afflatus. Few modern writers have been disposed to insist on the latter alternative, and have rather assumed that these references and details are accommodated to the state of knowledge at the time. I must observe here, however, that a careful consideration of the facts gives to a naturalist a much higher estimate of the real value of the observations of nature embodied in the Scriptures than that which divines have ordinarily entertained; and, consequently, that if we suppose them of human origin, we must be prepared to modify the views generally entertained of early Oriental simplicity and ignorance. The truth is, that a large proportion of the difficulties in Scriptural natural history appear to have arisen from want of such accommodation to the low state of the knowledge of nature among translators and expositors; and this is precisely what we should expect in a veritable revelation. Its moral and religious doctrines were slowly developed, each new light illuminating previous obscurities. Its human history comes out as evidence of its truth, when compared with monumental inscriptions; and why should not the All-wise have constructed as skilfully its teachings respecting his own works? There can be no doubt whatever that the Scripture writers intended to address themselves to the common mind, which now as then requires simple and popular teaching, but they were under obligation to give truthful statements; and we need not hesitate to say, with Dr. Chalmers, in reference to a book making such claims as those of the Bible : " There is no argument, saving that grounded on the usages of popular language, which would tempt us to meddle with the literalities of that ancient and, as appears to us, authoritative document, any farther than may be required by those conventionalities of speech which spring from 'optical' impressions of nature."*

Attempt as we may to disguise it, any other view is totally unworthy of the great Ruler of the universe, especially in a document characterized as emphatically *the truth*, and in a

---

* Much that is very silly has been written as to the extent of the supposed " optical view " taken by the Hebrew writers ; many worthy literary men appearing to suppose that *scientific* views of nature must necessarily be different from those which we obtain by the evidence of our senses. The very contrary is the fact ; and so long as any writers state correctly what they observe, without insisting on any fanciful hypotheses, science has no fault to find with them. What science most detests is the ignorant speculations of those who have not observed at all, or have observed imperfectly. It is a leading excellence of the Hebrew Scriptures that they state facts without giving any theories to account for them. It is, on the contrary, the circumstance that unscientific writers will not be content to be "optical," but must theorize, that spoils much of our modern literature, especially in its descriptions of nature.

moral revelation, in which statements respecting natural objects need not be inserted, unless they could be rendered at once truthful and illustrative of the higher objects of the revelation. The statement often so flippantly made that the Bible was not intended to teach natural history has no application here. *Spiritual* truths are no doubt shadowed forth in the Bible by material emblems, often but rudely resembling them, because the nature of human thought and language render this necessary, not only to the unlearned, but in some degree to all ; but this principle of adaptation can not be applied to plain material facts. Yet a confusion of these two very distinct cases appears to prevail almost unaccountably in the minds of many expositors. They tell us that the Scriptures ascribe bodily members to the immaterial God, and typify his spiritual procedure by outward emblems ; and this they think analogous to such doctrines as a solid firmament, a plane earth, and others of a like nature, which they ascribe to the sacred writers. We shall find that the writers of the Scriptures had themselves much clearer views, and that, even in poetical language, they take no such liberties with truth.

As an illustration of the extent to which this doctrine of "accommodation" carries us beyond the limits of fair interpretation, I cite the following passage from one of the ablest and most judicious writers on the subject : * "It was the opinion of the ancients that the earth, at a certain height, was surrounded by a transparent hollow sphere of solid matter, which they called the firmament. When rain descended, they supposed that it was through windows or holes made in the crystalline curtain suspended in mid-heavens. To these

---

* Prof. Hitchcock.

notions the language of the Bible is frequently conformed.
* * * But the most decisive example I have to give on this
subject is derived from astronomy.   Until the time of Coper-
nicus no opinion respecting natural phenomena was thought
better established than that the earth is fixed immovably in
the centre of the universe, and that the heavenly bodies move
diurnally round it.   To sustain this view the most decisive
language of Scripture might be quoted.   God is there said to
have '*established the foundations of the earth, so that they could
not be removed forever ;*' and the sacred writers expressly declare that the heavenly bodies *arise and set,* and nowhere allude to any proper motion of the earth."

   Will it be believed that, with the exception of the poetical
expression, " windows of heaven," and the common forms of
speech relating to sunrise and sunset, the above "decisive "
instances of accommodation have no foundation whatever in
the language of Scripture.   The doctrine of the rotation of
solid celestial spheres around the earth belongs to a Greek
philosophy which arose after the Hebrew cosmogony was
complete ; and though it occurs in the Septuagint and other
ancient versions, it is not based on the Hebrew original.   In
truth, we know that those Grecian philosophers—of the Ionic
and Pythagorean schools—who lived nearest the times of the
Hebrew writers, and who derived the elements of their
science from Egypt and Western Asia, taught very different
doctrines.   How absurd, then, is it thus to fasten upon the
sacred writers, contrary to their own words, the views of a
school of astronomy which probably arose long after their
time, when we know that more accurate ideas prevailed
nearer their epoch.   Secondly, though there is some reason
for stating that the "ancients," though certainly not those of
Israel, believed in celestial spheres supporting the heavenly

bodies, I suspect that the doctrine of a solid vault *supporting the clouds*, except as a mere poetical or mythological fancy, is a product of the imagination of the theologians and closet philosophers of a more modern time. The testimony of men's senses appears to be in favor of the whole universe revolving around a plane earth, though the oldest astronomical school with which we are acquainted suspected that this is an illusion; but the every-day observation of the most unlettered man who treads the fields and is wet with the mists and rains must convince him that there is no *sub-nubilar* solid sphere. If, therefore, the Bible had taught such a doctrine, it would have shocked the common-sense even of the plain husbandmen to whom it was addressed, and could have found no fit audience except among a portion of the literati of comparatively modern times. Thirdly, with respect to the foundations of the earth, I may remark that in the tenth verse of Genesis there occurs a definition as precise as that of any lexicon—"and God called the *dry land* earth;" consequently it is but fair to assume that the earth afterwards spoken of as supported above the waters is the dry land or continental masses of the earth, and no geologist can object to the statement that the dry land is supported above the waters by foundations or pillars.

We shall find in our examination of the document itself that all the instances of such accommodation which have been cited by writers on this subject are as baseless as those above referred to. It is much to be regretted that so many otherwise useful expositors have either wanted that familiarity with the aspects of external nature by which all the Hebrew writers are characterized, or have taken too little pains to ascertain the actual meaning of the references to creation which they find in the Bible. I may further remark that if such instances of accommodation could be found in the later poetical books, it would be extremely unfair to apply them as aids in the interpretation of the plain, precise, and unadorned statements of the first chapters of Genesis. There is, however, throughout even the higher poetry of the Bible, a truthful representation and high appreciation of nature for which we seek in vain in any other poetry, and we may fairly trace this in part to the influence of the cosmogony which appears in its first chapter. The Hebrew was thus taught to recognize the unity of nature as the work of an Almighty Intelligence, to regard all its operations as regulated by his unchanging law or " decree," and to venerate it as a revelation of his supreme wisdom and goodness. On this account he was likely to regard careful observation and representation with as scrupulous attention as the modern naturalist. Nor must we forget that the Old Testament literature has descended to us through two dark ages—that of Greek and Roman polytheism and of Middle Age barbarism—and that we must not confound its tenets with those of either. The religious ideas of both these ages were favorable to certain forms of literature and art, but eminently unfavorable to the successful prosecution of the study of nature. Hence we have a right to expect in the literature of the golden age of primeval monotheism more affinity with the ideas of modern science than in any intermediate time; and the truthful delineation which the claims of the Bible to inspiration require might have been, as already hinted, to a certain extent secured merely by the reflex influence of its earlier statements, without the necessity of our supposing that illustrations of this kind in the later books came directly from the Spirit of God.

Our discussion of this part of the subject has necessarily been rather desultory, and the arguments adduced must de-

pend for their full confirmation on the results of our future
inquiries. The conclusions arrived at may be summed up as
follows : 1. That the Mosaic cosmogony must be considered,
like the prophecies of the Bible, to claim the rank of inspired
teaching, and must depend for its authority on the maintenance
of that claim. 2. That the incidental references to nature in
other parts of Scripture indicate, at least, the influence of
these earlier teachings, and of a pure monotheistic faith, in
creating a high and just appreciation of nature among the
Hebrew people.

It is now necessary to inquire in what precise form this remarkable revelation of the origin of the world has been given.
I have already referred to the hypothesis that it represents a
vision of creation presented to the mind of a seer, as if in a
series of pictures which he represents to us in words. This
is perhaps the most intelligible conception of the manner of
communication of a revelation from God ; and inasmuch as it
is that referred to in other parts of the Bible as the mode of
presentation of the future to inspired prophets, there can be
no impropriety in supposing it to have been the means of
communicating the knowledge of the unknown past. We
may imagine the seer — perhaps some aboriginal patriarch,
long before the time of Moses—perhaps the first man himself
—wrapt in ecstatic vision, having his senses closed to all the
impressions of the present time, and looking as at a moving
procession of the events of the earth's past history, presented
to him in a series of apparent days and nights. In the first
chapter of Genesis he rehearses this divine vision to us, not
in poetry, but in a series of regularly arranged parts or
strophes, thrown into a sort of rhythmical order fitted to impress them on the memory, and to allow them to be handed
down from mouth to mouth, perhaps through successive gen-

erations of men, before they could be fixed in a written form of words. Though the style can scarcely be called poetical, since its expressions are obviously literal and unadorned by figures of speech, the production may not unfairly be called the Song or Ballad of Creation, and it presents an Archaic simplicity reminding us of the compositions of the oldest and rudest times, while it has also an artificial and orderly arrangement, much obscured by its division into verses and chapters in our Bibles. It is undoubtedly also characterized by a clearness and grandeur of expression very striking and majestic, and which shows that it was written by and intended for men of no mean and contracted minds, but who could grasp the great problems of the origin of things, and comprehend and express them in a bold and vigorous manner. It may be well, before proceeding farther, to present to the reader this ancient document in a form more literal and intelligible, and probably nearer to its original dress, than that in which we are most familiar with it in our English Bibles:

THE ABORIGINAL SONG OF CREATION.

Beginning.

 In the Beginning God created the Heavens and the Earth,
 And the Earth was formless and empty,
 And darkness on the surface of the deep,
 And the Breath of God moved on the Surface of the Waters.

Day One.

 And God said—" Let Light be,"
 And Light was.
 And God saw the Light that it was good.
 And God called the Light Day,
 And the darkness he called Night.
 And Evening was and Morning was—Day one.

Day Second.

 And God said—" Let there be an Expanse in the midst of the waters,

And let it divide the waters from the waters."
And God made the Expanse,
And divided the waters below the Expanse from the waters above
the Expanse.
And it was so.
And God called the Expanse Heavens.
And Evening was and Morning was, a Second Day.

Day Third.
And God said—" Let the waters under the Heavens be gathered into
one place,
And let the Dry Land appear."
And it was so,
And God called the Dry Land Earth,
And the gathering of waters called he Seas.
And God saw that it was good.
And God said—" Let the earth shoot forth herbage,
The Herb yielding seed and the fruit-tree yielding fruit containing
seed after its kind, on the earth."
And it was so.
And the earth brought forth herbage,
The Herb yielding seed and the Tree yielding fruit whose seed is in
it after its kind,
And God saw that it was good.
And Evening was and Morning was, a Third Day.

Day Fourth.
And God said—" Let there be Luminaries in the Expanse of Heaven,
To divide the day from the night,
And let them be for Signs and for Seasons,
And for Days and for Years.
And let them be Luminaries in the Expanse of Heaven
To give light on the earth."
And it was so.
And God made two great Luminaries,
The greater Luminary to rule the day,
The lesser Luminary to rule the night,
The Stars also.
And God placed them in the Expanse of Heaven

To give light upon the earth,
And to rule over the day and over the night,
And to divide the light from the darkness.
And God saw that it was good.
And Evening was and Morning was, a Fourth Day.

Day Fifth.

And God said—" Let the waters swarm with swarmers, having life,
And,let winged animals fly over the earth on the surface of the expanse of heaven."
And God created great Reptiles,
And every living thing that moveth,
With which the waters swarmed after their kind,
And every winged bird after its kind.
And God saw that it was good.
And God blessed them, saying—
" Be fruitful and multiply,
And fill the waters of the sea ;
And let birds multiply in the land."
And Evening was and Morning was, a Fifth Day.

Day Sixth.

And God said—" Let the Land bring forth living things after their kind,
Herbivores and smaller mammals and Carnivores after their kind."
And it was so.
And God made all Carnivores after their kind,
And all Herbivores after their kind,
And all minor mammals after their kind.
And God saw that it was good.
And God said—" Let us make man in our image, after our likeness,
And let him have dominion over the fish in the sea
And over the birds of the heavens,
And over the Herbivora,
And over the Earth,
And over all the minor animals that creep upon the earth."
And God created man in his own image,
In the image of God created he him,
Male and female created he them.

And God blessed them.

And God said unto them—

" Be fruitful and multiply,

And replenish the earth and subdue it,

And have dominion over the fishes of the sea

And over the birds of the air,

And over all the animals that move upon the earth."

And God said—" Behold, I have given you all herbs yielding seed,

Which are on the surface of the whole earth,

And every tree with fruit having seed,

They shall be unto you for food.

And to all the animals of the land

And to all the birds of the heavens,

And to all things moving on the land having the breath of life,

I have given every green herb for food."

And it was so.

And God saw every thing that he had made, and behold it was very
good.

And Evening was and Morning was, a Sixth Day.

Day Seventh.

Thus the Heavens and the Earth were finished,

And all the hosts of them.

And on the seventh day God ended the work which he had made,

And he rested on the seventh day from all his work which he had
made.

And God blessed the seventh day and hallowed it,

Because that in it he rested from all his work that he had created and
made.

CHAPTER III.

> " What if earth
> Be but a shadow of heaven, and things therein
> Each to the other like ; more than on earth is thought."
>
> MILTON.

(3) *Character of the Biblical Cosmogony, and general Views of Nature which it Contains or to which it Leads.*—Much of what appertains to the character of the revelation of origins has been anticipated under previous heads. We have only to read the Song of Creation, as given in the last chapter, to understand its power and influence as a beginning of religious doctrine. The revelation was written for plain men in the infancy of the world. Imagine Chaldean or Hebrew shepherd listening to these majestic lines from the lips of some ancient patriarch, and receiving them as truly the words of God. What a grand opening to him of both the seen and unseen worlds! Henceforth he has no superstitious dread of the stars above, or of the lightning and thunder, or of the dark woods and flowing waters beneath. They are all the works of the one Creator, the same Creator who is his own Maker, in whose image and shadow he is made. He can look up now to the heavens or around upon the earth, and see in all the handiwork of God, and can worship God through all. He can see that the power that cares

for the birds and the flowers of the field cares for him. He is no longer the slave and sport of unknown and dreadful powers ; they are God's workmanship and under his control —nay, God has given him a mission to subdue and rule over them. So these noble words raise him to a new manhood, and emancipate him from the torture of endless fears, and open to him vast new fields of thought and inquiry, which may enrich him with boundless treasures of new religious and intellectual wealth. Imagine still farther that he wanders into those great cities which are the seats of the idolatries of his time. He enters magnificent temples, sees elaborately decorated altars, huge images, gorgeous ceremonials, priests gay in vestments and imposing in numbers. He is invited to bow down before the bull Apis, to worship the statue of Belus or of Ishtar, of Osiris or of Isis. But this is not in his book of origins. All these things are contrivances of man, not works of God, and their aim is to invite him to adore that which is merely his fellow-creature, that which he has the divine commission to subdue and rule. So our primitive Puritan turns away. He will rather raise an altar of rough stones in the desert, and worship the unseen yet real Creator, the God that has no local habitation in temples made with hands, yet is everywhere present. Such is the moral elevation to which this revelation of origins raises humanity ; and when there was added to it the farther history of primeval innocence, of the fall, and of the promise of a Redeemer, and of the fate of the godless antediluvians, there was a whole system of religion, pure and elevating, and placing the Abrahamidæ, who for ages seem alone to have held to it, on a plane of spiritual vantage immeasurably above that of other nations. Farther, every succeeding prophet whose works are included in the sacred canon, following up these doctrines in the same spirit, and added new treasures of divine knowledge from age to age.

But admitting all this, it may be asked, Are these ancient records of any value to us? May we not now dispense with them, and trust to the light of science? The infinitely varied and discordant notions of our modern literature on these great questions of origin, the incapacity of any philosophical system to reach the common mind for practical purposes, and the baseless character of any religious system which does not build on these great primitive truths, give a sufficient answer. Farther, we may affirm that the greatest and widest generalizations of our modern science have, in so far as they are of practical importance, been anticipated in the revelations of the Bible, and that in the cosmogony of Genesis and its continuation in the other sacred books we have general views of the universe as broad as those of any philosophies, ancient or modern. This is a hard test for our revelation, but it can be endured, and we may shortly inquire what we find in the Bible of such great general truths.

Many may be disposed to admit the accurate delineation of natural facts open to human observation in the sacred Scriptures, who may not be prepared to find in these ancient books any general views akin to those of the ancient philosophers, or to those obtained by inductive processes in modern times. Yet views of this kind are scattered through the Hebrew and Christian Scriptures, and are a natural outgrowth and development of the great facts and principles asserted in the first chapter of Genesis. They resolve themselves, almost as a matter of course, into the two leading ideas of order and adaptation. I have already quoted the eloquent admission by Baron Humboldt of the presence of these ideas of the cosmos in Psalm civ. They are both conspicuous in the narrative of creation, and equally so in a great number of other passages. ' "Order is heaven's first law; and the second is like unto it—that every thing serves an end. This is the sum of all science. These are the two mites, even all that she hath, which she throws into the treasury of the Lord; and, as she does so in faith, Eternal Wisdom looks on and approves the deed."* These two mites, lawfully acquired by science, by her independent exertions, she may, however, recognize as of the same coinage with the treasure already laid up in the rich storehouse of the Hebrew literature; but in a peculiar and complex form, which may be illustrated under the following general statements :

1. The Scriptures assert invariable natural law, and constantly recurring cycles in nature. Natural law is expressed as the ordinance or decree of Jehovah. From the oldest of the Hebrew books I select the following examples : †

> "When he made a decree for the rain,
> And a way for the thunder-flash."
> —Job xxviii., 26.

> "Knowest thou the ordinances of the heavens ?
> Canst thou establish a dominion even over the earth ?"
> —Job xxxviii., 33.

The later books give us such views as the following :

> "He hath established them [the heavens] for ever and ever ;
> He hath made a decree which shall not pass."
> —Psa. cxlviii., 6.

---

* McCosh, "Typical Forms and Special Ends."
† I adopt that view of the date of Job which makes it precede the Exodus, because the religious ideas of the book are patriarchal, and it contains no allusions to the Hebrew history or institutions. Were I to suggest an hypothesis as to its origin, it would be that it was written or found by

"Thou art forever, O Jehovah, thy word is established in the heavens;
   Thou hast established the earth, and it abideth;
   They continue this day according to thine ordinances, for all are thy
      servants."
                                                          —Psa. cxix., 90.

      "When he established the clouds above;
      When he strengthened the fountains of the deep;
      When he gave to the sea his decree,
      That the waters should not pass his commandment;
      When he appointed the foundations of the earth."
                                                          —Prov. viii., 28.

Many similar instances will be found in succeeding pages;
and in the mean time we may turn to the idea of recurring
cycles, which forms the starting-point of the reasonings of
Solomon on the current of human affairs, in the book of
Ecclesiastes: "One generation passeth away, and another
generation cometh; but the earth abideth for the ages. The
sun ariseth, and the sun goeth down, and hasteneth to its
place whence it arose. The wind goeth toward the south,
and turneth unto the north. It whirleth about continually,
and returneth again according to its circuits. All the rivers
run into the sea, yet the sea doth not overflow; unto the
place whence the rivers came, thither they return again." I
might fill pages with quotations more or less illustrative of
the statement in proof of which the above texts are cited; but
enough has been given to show that the doctrine of the Bible
is not that of fortuitous occurrence, or of materialism, or of
pantheism, or of arbitrary supernaturalism, but of invariable
natural law representing the decree of a wise and unchanging
Creator. It is a common but groundless and shallow charge
against the Bible that it teaches an "arbitrary supernatural-

---

Moses when in exile, and published among his countrymen in Egypt, to
revive their monotheistic religion, and cheer them under the apparent de-
sertion of their God and the evils of their bondage.

ism." ✝ What it does teach is that all nature is regulated by
the laws of God, which like himself are unchanging, but which
are so complex in their relations and adjustments that they
allow of infinite variety, and do not exclude even miraculous
intervention, or what appears to our limited intelligence as
such. ⏐ In opposition to this, it is true, some physicists have
held that natural law is a fatal necessity.* If they mean by
this a merely hypothetical necessity that certain effects must
follow if certain laws act, this is in accordance with the Biblical view, for nothing can resist the will of God. But if they
mean an absolute necessity that these laws can not be suspended or counteracted by higher laws, or by the will of the
Creator, they assert what is not only contrary to Scripture, but
absurd, for " blind metaphysical necessity, which is the same
always and everywhere, could produce no variety of things."✝
It could lead merely to a dead and inert equilibrium. On the
hypothesis of mere physical necessity, the universe either never
could have existed, or must have come to an end infinite ages
ago, which is the same thing. Only on the hypothesis of law
proceeding from an intelligent will can we logically account
for nature.

2. The Bible recognizes progress and development in nature. At the very outset we have this idea embodied in the
gradual elaboration of all things in the six creative periods,
rising from the formless void of the beginning, through successive stages of inorganic and organic being, up to Eden
and to man. Beyond this point the work of creation stops;
but there is to be an occupation and improvement of the
whole earth by man spreading from Eden. This process is
arrested or impeded by sin and the fall. Here commences

---

* Tyndall seems to hold this.   † Newton.

the special province of the Bible, in explaining the means of
recovery from the fall, and of the establishment of a new
spiritual and moral kingdom, and finally of the restoration of
Eden in a new heaven and earth.   All this is moral, and relates to man, in so far as the present state of things is concerned ; but we have the commentary of Jesus : " My Father
worketh hitherto, and I work ;" the remarkable statement of
Paul, that the whole creation is involved in the results of
man's moral fall and restoration, and the equally remarkable
one that the Redeemer is also the maker of the " worlds " or
ages of the earth's physical progress, as well as of the future
" new heaven and new earth."   Peter also rebukes indignant-
ly those scoffers who maintained that all things had remained
as they are since the beginning ; and refers to the creation
week and to the deluge as earnests of the great changes yet
in store for the earth.*

It is indeed curious to observe how in our version of the
Bible this idea of progress in the universe, or of " time-worlds,"
as it has been called, has been variously replaced by the words
" world " and " eternity," owing to the defective ideas prevalent at the time when the translation was made.   In the Hebrew Scriptures the term *Olam*, " age," and in the New Testament the equivalent term *Aiōn* have been thus treated, and
their real significance much obscured.   Thus when it is said,
" by faith we understand that the *worlds* were framed," or " by
him God made the *worlds*,"† or that certain of God's plans
have been hid " from the beginning of the *world*,"‡ the reference is not to worlds in space, but to worlds in time, or ages
of God's working in the universe.   So also these ages of God's

---

* John v., 17 ; Rom. viii., 22 ; Heb. i., 2 ; 2 Peter iii.
† Heb. i., 2.   ‡ Eph. iii., 9.

working are given to us as our only intelligible type of eternity, of which absolutely we can have no conception. Thus God's "eternal purpose" is his purpose of the ages. So when he is the "King eternal,"* and in that capacity gives to his people "life everlasting," he is the King of the ages, and gives life of the ages. So in the noble hymn attributed to Moses (Psalm xc.), where our version has, "from everlasting to everlasting thou art God," the original is, "from age to age thou art, O God." It has perhaps been a defect of our modern science that it has familiarized us merely with the existence of worlds in space, and not with their existence in time. It is only in comparatively modern times that the developments of chronological geology and of physical astronomy have brought before us, not only the long ages in which the earth was passing through its formative stages, but also the fact that still longer æons are embraced in the history of the other bodies of our solar system, and of the starry orbs and nebulæ. These grand conceptions were already embodied in the Hebrew revelation, and were used there as the means of giving some faint approach to a conception of the unlimited existence of God himself, of the ages in which his creative work has been going on, and of the future life he has prepared for his redeemed people.

Such views of development and progress are not unknown to many ancient cosmogonies and philosophical systems, but they had no stable foundation in observed fact until the rise of modern geology and physical astronomy; which enable us to affirm that, in addition to those changeless physical laws which cause the bodies of the universe to wheel in unvarying cycles, and all natural powers to reproduce themselves, and, in addition to those organic laws which produce unceasing successions of living individuals, there is a higher law of progress. We can now trace back man, the animals and plants his contemporaries, and others which preceded them, our continents and mountain ranges, and the solid rocks of which they are composed—nay, the very fabric of the solar system itself—to their several origins at distinct points of time; and can maintain that since the earth began to wheel around the sun, no succeeding year has seen it precisely as it was in the year before. The old Hebrew record affirms, and I presume scarcely any sane man really doubts, that this law of progress emanates from the mind and power of one creative Being. When men see in natural law only recurring cycles, they may be pardoned for falling even into the absurdity of believing in eternal succession; but when they see change and progress, and this in a uniform direction, overmastering recurring cycles, and introducing new objects and powers not accounted for by previous objects or powers, they are brought very near to the presence of the Spiritual Creator. And hence, although no science can reach back to the act of creation, this doctrine is much more strongly held in our day by geologists than by physicists. It is quite true that the idea of creative acts has been superseded to a great extent by that of "creation by law," or by that of "evolution." Still behind all there lies a primary creative power; and the validity of these ideas and their bearing on theism and creation we shall have to discuss in the sequel. In one thing only does the Bible here part company with natural science. The Bible goes on into the future, and predicts a final condition of our planet, of which science can from its investigations learn nothing.

3. The Bible recognizes purpose, use, and special adapta-

tion in nature. It is, in short, full of natural theology, akin in some respects to that which has been so elaborately worked out by so many modern writers. Numerous passages in support of this will occur to every one who has read the Scriptures. It is necessary here, however, to direct attention to a distinction very obvious in Scripture, but not always attended to by writers on this subject. The Bible maintains the true "final cause" of all nature to be, not its material and special adaptations or its value to man, but the pleasure or satisfaction of the Creator himself. In the earlier periods of Creation, before man was upon the earth, God contemplates his work and pronounces it good. The heavenly hosts praise him, saying, " Thou hast created all things, and for thy pleasure they are and were created." Further, the Bible represents intelligences higher than man as sharing in the delight which may be derived from the contemplation of God's works. When the earth first rose from the waters to greet the light, " the morning stars sang together, and all the sons of God shouted for joy." There are many things in nature that strongly impress the naturalist with this same view, that the Creator takes pleasure in his works; and, like human genius in its highest efforts, rejoices in production, even if no sentient being should be ready to sympathize. The elaborate structures of fossils, of which we have only fragmentary remains, the profusion of natural objects of surpassing beauty that grow and perish unseen by us, the delicate microscopic mechanism of nearly all organic structures, point to other reasons for beauty and order than those that concern man, or the mere utilities of human beings; and though there are now naturalists who deny absolutely that beauty is an object in nature, and assign even the colors of flowers and insects to utility alone, and this of a very low order, this doctrine is so

repulsive to our higher sentiments that there is little danger
of its general acceptance; while the slightest consideration
shows that the utilities referred to could have been secured
without any of this consummate beauty associated with them,
and our perception of and delight in which mark in a way
beyond the ability of skepticism to cavil at our own spiritual
kinship with the Author of all this profusion of beauty. Yet
man is represented as the chief created being for whom this
earth has been prepared and designed. He obtains dominion
over it. A chosen spot is prepared for him, in which not only
his wants but his tastes are consulted; and, being made in
the image of his Maker, his æsthetic sentiments correspond
with the beauties of the Maker's work, and he finds there
also food for his reason and imagination. This view of the
subject, as well as others already referred to, is finely represented in the address of the Almighty to Job.*

The Bible also very often refers to the special adaptations of natural objects and laws to each other, and to the
promotion of the happiness of sentient creatures lower than
man. The 104th Psalm is replete with notices of such
adaptations, and so is the address to Job; and indeed this
view seems hardly ever absent from the minds of the Hebrew
writers, but has its highest applications in the lilies of the
field, that toil not neither do they spin, and the sparrows that
are sold for a farthing, yet the heavenly Father has clothed
the one with surpassing beauty, and provides food for the
other, nor allows it to fall without his knowledge. I may, by
way of farther illustration, merely name a few of the adaptations referred to in Job xxxviii. and the following chapters.
The winds and the clouds are so arranged as to afford the

* Job xxxviii. and xxxix.

required supplies of moisture to the wilderness where no man is, to "cause the bud of the tender herb to spring forth." For similar objects the tempest is ordered, and the clouds arranged "by wisdom." The adaptations of the wild ass, the wild goat, the ostrich, the migratory birds, the horse, the hippopotamus, the crocodile, to their several habitats, modes of life, and uses in nature, are most vividly sketched and applied as illustrations of the consummate wisdom of the Creator, which descends to the minutest details of organization and habit.

It is to be observed here that in holding this doctrine of use and adaptation in nature, the Bible is only consistent with its own theory of rational theism. The Monotheist can not refer nature to a conflict of antagonistic powers and forces. He must recognize in it a unity of plan ; and even those things which appear aberrant, irregular, or noxious must have their place in this plan. Hence in the Bible God is maker not only of the day but of the night, not only of the peaceful cattle but of the voracious crocodile, not only of the sunshine and shower but of the tornado and the earthquake. Further, in all these things God is manifested, so that we may learn "his eternal power and divinity* from the things which he has made," and in all these also there are emblems of his relations to us. This argument from design is in truth the only proof the Bible condescends to urge for the existence of God ; and it is the only one in which in his later days our great English philosopher Mill could see any validity.†

If the reader happens to be familiar with the objections to the doctrine of final causes, or teleology, in nature, urged in

---

* Romans i., 20.　　　　　　　† Essays on Theism.

our day by Spencer, Haeckel, and others, he will have seen
from the foregoing statements that these objections are in
themselves baseless, or inapplicable to this doctrine as maintained in the Bible.   There is no consistency in the position
of men who, when they dig a rudely chipped flint out of a bed
of gravel, immediately infer an intelligent workman, and who
refuse to see any indication of a higher intelligence in the
creation of the workman himself.   It is a blind philosophy
which professes to see in primal atoms the "promise and
potency of mind," and which fails to perceive that such
potency is more inconceivable than the evidence of primary
and supreme mind.   The men who maintain that wings were
not planned for flight, but that flight has produced wings,
and thousands of like propositions, are simply amusing themselves with paradoxes to which may very properly be applied
the strange word devised by Haeckel to express his theory of
nature—*Dysteleology*, or purposelessness.   It is to be borne
in mind, however, that the teleology of the Bible is not of that
narrow kind which would make man the sole object of nature, and the supreme judge of its adaptations.   Inasmuch as
God's plan goes over all the ages past and future, and relates
to the welfare of all sentient beings known or unknown to us,
and also to his own sovereign pleasure as the supreme object,
we may not be in a position either to understand or profit by
all its parts, and hence may expect to find many mysteries,
and many things that we can not at present reconcile with
God's wisdom and goodness.   We know but "parts of his
ways," the "fullness of his power who can understand."
"His judgments are unsearchable," "his ways are past
finding out."

4. The law of type or pattern in nature is distinctly indicated in the Bible.   This is a principle only recently understood

by naturalists, but it has more or less dimly dawned on the
minds of many great thinkers in all ages. Nor is this wonderful, for the idea of type is scarcely ever absent from our
own conceptions of any work that we may undertake. In
any such work we anticipate recurring daily toil, like the returning cycles of nature. We look for progress, like that of
the growth of the universe. We study adaptation both of the
several parts to subordinate uses, and of the whole to some
general design. But we also keep in view some pattern, style,
or order, according to which the whole is arranged, and the
mutual relations of the parts are adjusted. The architect
must adhere to some order of architecture, and to some style
within that order. The potter, the calico-printer, and the
silversmith must equally study uniformity of pattern in their
several manufactures. The Almighty Worker has exhibited
the same idea in his works. In the animal kingdom, for instance, we have four or more leading types of structure.
Taking any one of these—the vertebrate, for example—we
have a uniform general plan, embracing the vertebral column
constructed of the same elements ; the members, whether the
arm of man, the limb of the quadruped, or the wing of the
bat or the bird, or the swimming-paddle of the whale, built of
the same bones. In like manner all the parts of the vertebral column itself in the same animal, whether in the skull,
the neck, or the trunk, are composed of the same elementary
structures. These types are farther found to be sketched out
—first in their more general, and then in their special features
—in proceeding from the lower species of the same type to
the higher, in proceeding from the earlier to the later stages
of embryonic development, and in proceeding from the more
ancient to the more recent creatures that have succeeded
each other in geological time. Man, the highest of the ver-

tebrates, is thus the archetype, representing and including all
the lower and earlier members of the vertebrate type.   The
above are but trite and familiar examples of a doctrine which
may furnish and has furnished the material of volumes.
There can be no question that the Hebrew Bible is the oldest book in which this principle is stated.   In the first chapter of Genesis we have specific type in the creation of plants
and animals after their kinds or species, and in the formation
of man in the image and likeness of the Creator ; and, as we
shall find in the sequel, there are some curious ideas of higher and more general types in the grouping of the creatures
referred to.   The same idea is indicated in the closing chapters of Job, where the three higher classes of the vertebrates
are represented by a number of examples, and the typical
likeness of one of these—the hippopotamus—to man, seems
to be recognized.   Dr. McCosh has quoted, as an illustration
of the doctrine of types, a very remarkable passage from
Psalm cxxxix. :

> "I will praise Thee, for I am fearfully and wonderfully made.
> Marvellous are thy works,
> And that my soul knoweth right well.
> My substance was not hid from Thee,
> When I was made in secret,
> And curiously wrought in the lowest parts of the earth :
> Thine eyes did see my substance, yet being imperfect ;
> And in thy book all my members were written,
> Which in continuance were fashioned when as yet there was none
>   of them."

It would too much tax the faith of many to ask them to
believe that the writer of the above passage, or the Spirit that
inspired him, actually meant to teach—what we now know so
well from geology—that the prototypes of all the parts of the

archetypal human structure may be found in those fossil remains of extinct animals which may, in nearly every country,
be dug up from the rocks of the earth. No objection need,
however, be taken to our reading in it the doctrine of embryonic development according to a systematic type.

Science, it is true, or rather I should perhaps say philosophical speculation, has sometimes pushed this idea of plan
into that of a spontaneous genetic evolution of things in time,
without any creative superintendence or definite purpose.
This way of viewing the matter is, however, as we shall have
occasion to see, both bald and irrational, and wants the symmetry and completeness of that style of thought which grasps
at once progress and plan and adaptation, as emanating from
a Supreme Will. The question of how the plan has been
worked out will come up for detailed consideration farther
on. In the mean time we have before us the fact that the
Bible represents the cosmos as not the product of a blind
conflict of self-existent forces, but as the result of the production and guidance of these forces by infinite wisdom.

It is more than curious that this idea of type, so long existing in an isolated and often depised form, as a theological
thought in the imagery of Scripture, should now be a leading idea of natural science ; and that while comparative
anatomy teaches us that the structures of all past and present lower animals point to man, who, as Professor Owen expresses it, has had all his parts and organs " sketched out in
anticipation in the inferior animals," the Bible points still
farther forward to an exaltation of the human type itself into
what even the comparative anatomist might perhaps regard
as among the " possible modifications of it beyond those
realized in this little orb of ours," could he but learn its real
nature.

Under the foregoing heads, of the object, the structure, the authority, and the general cosmical views of the Scripture, I have endeavored to group certain leading thoughts important as preliminary to the study of the subject ; and, in now entering on the details of the Old Testament cosmogony, I trust the reader will pardon me for assuming, as a working hypothesis, that we are studying an inspired book, revealing the origin of nature, and presenting accurate pictures of natural facts and broad general views of the cosmos, at least until in the progress of our inquiry we find reason to adopt lower views ; and that he will, in the mean time, be content to follow me in that careful and systematic analysis which a work claiming such a character surely demands.

CHAPTER IV.

THE BEGINNING.

"In the beginning Elohim created the heavens and the earth."—Genesis i., 1.

IT is a remarkable and instructive fact that the first verse of the Hebrew sacred writings speaks of the material universe—speaks of it as a whole, and as originating in a power outside of itself. The universe, then, in the conception of this ancient writer, is not eternal. It had a beginning, but that beginning in the indefinite and by us unmeasured past. It did not originate fortuitously, or by any merely accidental conflict of self-existent material atoms, but by an act—an act of will on the part of a Being designated by that name which among all the Semitic peoples represented the ultimate, eternal, inscrutable source of power and object of awe and veneration. With the simplicity and child-like faith of an archaic age, the writer makes no attempt to combat any objections or difficulties with which this great fundamental truth may be assailed. He feels its axiomatic force as the basis of all true religion and sound philosophy, and the ultimate fact which must ever bar our further progress in the investigation of the origin of things—the production from non-existence of the material universe by the eternal self-existent God.

It did not concern him to know what might be the nature of that unconditioned self - existence ; for though, like our ideas of space and time, incomprehensible, it must be assumed. It did not concern him to know how matter and force subsist, or what may be the difference between a material universe cognizable by our senses and the absolute want of all the phenomena of such a universe or of whatever may be their basis and essence. Such questions can never be answered, yet the succession of these phenomena must have had a commencement somewhere in time. How simple and how grand is his statement! How plain and yet how profound its teachings!

It is evident that the writer grasps firmly the essence of the question as to the beginning of things, and covers the whole ground which advanced scientific or philosophical speculation can yet traverse. That the universe must have had a beginning no one now needs to be told. If any philosophical speculator ever truly held that there has been an endless succession of phenomena, science has now completely negatived the idea by showing us the beginning of all things that we know in the present universe, and by establishing the strongest probabilities that even its ultimate atoms could not have been eternal. But the question remains—If there was a beginning, what existed in that beginning? To this question many partial and imperfect answers have been given, but our ancient record includes them all.

If any one should say, "In the beginning was nothing." Yes, says Genesis, there was, it is true, nothing of the present matter and arrangements of nature. Yet all was present potentially in the will of the Creator.

"In the beginning were atoms," says another. Yes, says Genesis, but they were created; and so says modern science, and must say of ultimate particles determined by weight and measure, and incapable of modification in their essential properties—" They have the properties of a manufactured article." *

" In the beginning were forces," says yet another. True, says Genesis; but all forces are one in origin—they represent merely the fiat of the eternal and self-existent. So says science, that force must in the ultimate resort be an " expression of Will." †

" In the beginning was Elohim," adds our old Semitic authority, and in him are the absolute and eternal thought and will, the Creator from whom and by whom and in whom are all things.

Thus the simple familiar words, " In the beginning God created the heaven and the earth," answer all possible questions as to the origin of things, and include all under the conception of theism. Let us now look at these pregnant words more particularly as to their precise import and significance.

The divine personality expressed by the Hebrew Elohim may be fairly said to include all that can be claimed for the pantheistic conception of "dynamis," or universal material power. Lange gives this as included in the term Elohim, in his discussion of this term in his book on Genesis. It has been aptly said that if, physically speaking, the fall of a sparrow produces a gravitative effect that extends throughout the universe, there can be no reason why it should be unknown to God. God is thus everywhere, and always. Yet he is everywhere and always present as a personality knowing and willing. From his thought and will in the beginning proceeded the universe. By him it was created.

---

* Herschel, Dissertation on the Study of Natural Philosophy ; Maxwell, Lecture before the British Association.
† Carpenter, " Human Physiology."

What, then, is creation in the sense of the Hebrew writer. The act is expressed by the verb *bara*, a word of comparatively rare occurrence in the Scriptures, and employed to denote absolute creation, though its primary sense is to cut or carve, and it is indeed a near relative of our own English word "pare." If, says Professor Stuart, of Andover, this word " does not mean to create in the highest sense, then the Hebrews had no word by which they could designate this idea." Yet, like our English "create," the word is used in secondary and figurative senses, which in no degree detract from its force when strictly and literally used. Since, however, these secondary senses may often appear to obscure the primitive meaning, we must examine them in detail.

In the first chapter of Genesis, after the general statement in verse 1, other verbs signifying to *form* or *make* are used to denote the elaboration of the separate parts of the universe, and the word "create" is found in only two places, when it refers to the introduction of "great whales" (reptiles) and of man. These uses of the word have been cited to disprove its sense of absolute creation. It must be observed, however, that in the first of these cases we have the earliest appearance of animal life, and in the second the introduction of a rational and spiritual nature. Nothing but pure materialism can suppose that the elements of vital and spiritual being were included in the matter of the heavens and the earth as produced in the beginning ; and as the Scripture writers were not materialists, we may infer that they recognized, in the introduction of life and reason, acts of absolute creation, just as in the origin of matter itself. In Genesis ii. and iii. we have a form of expression which well marks the distinction between creation and making. God is there said to have rested from all his works which he "created and made"

—literally, created " for or in reference to making," the word for making being one of those already referred to.* The force of this expression consists in its intimating that God had not only finished the work of *creation*, properly so called, but also the elaboration of the various details of the universe, as formed or fashioned out of the original materials. Of a similar character is the expression in Isaiah xlii., 5, " Jehovah, he that *created* the heavens and spread them out ;" and that in Psalm cxlviii., 5, " He commanded and they were *created*, he hath also established them for ever and ever."

In as far as I am aware, the word *bara* in all the remaining instances of its occurrence in the Pentateuch refers to the creation of man, with the following exceptions : Exodus xxxiv., 10, " I will do (create) marvels, such as have not been seen in all the earth ;" Numbers xvi., 30, " If the Lord make a new thing (create a creation), and the earth open her mouth and swallow them up." These verses are types of a class of expressions in which the proper term for creation is applied to the production of something new, strange, and marvellous ; for instance, " Create in me a clean heart, O Lord ;" "Behold, I create new heavens and a new earth." It is, however, evidently an inversion of sound exposition to say that these secondary or figurative meanings should determine the primary and literal sense in Genesis i. On the contrary, we should rather infer that the sacred writers in these cases selected the proper word for creation, to express in the most forcible manner the novel and thorough character of the changes to which they refer, and their direct dependence on the Divine will. By such expressions we are in effect referred back to the original use

* Asah.

of the word, as denoting the actual creation of matter by the
command of God, in contradistinction from those arrangements which have been effected by the gradual operation of
secondary agents, or of laws attached to matter at its creation.   It has been farther observed* that in the Hebrew
Scriptures this word *bara* is applied to God only as an
agent, not to any human artificer ; a fact which is very important with reference to its true significance.   Viewing creation in this light, we need not perplex ourselves with the
question whether we should consider Genesis i., 1, to refer
to the essence of matter as distinguished from its qualities.
We may content ourselves with the explanation given by
Paul in the eleventh of Hebrews : " By faith we are certain
that the worlds† were created by the decree of God, so that
that which *is seen* was made of that which *appears not."*   Or,
with reference to the other uses of the word, if the first introduction of animal life was a creation, and if the introduction of the rational nature of man was a creation, we may
suppose that the original creation was in like manner the introduction or first production of those entities which we call
matter and force, and which to science now are as much ultimate facts as they were to Moses.

The *nature* of the act of creation being thus settled, its
extent may be ascertained by an examination of the terms
heaven and earth.

The word "heavens" (*shamayim*) has in Hebrew as in
English a variety of significations.   Of material heavens there
are, in the quaint language of Poole, "*tres regiones, ubi aves,
ubi nubes, ubi sidera;*" or (1) the atmosphere or firmament ;‡

* McDonald, "Creation and the Fall."
† Literally, "ages" or "time-worlds," as they have been called.
‡ Genesis i., 8, 26–28.

(2) the region of clouds in the upper part of the atmosphere;*
(3) the depths of space comprehending the starry orbs.† Besides these we have the "heaven of heavens," the abode of God
and spiritual beings.‡ The application of the term "heav-
en" to the atmosphere will be considered when we reach the
6th and 7th verses. In the mean time we may accept the
word in this place as including the material heavens in the
widest sense : (1.) Because it is not here, as in verse 8th, restricted to the atmosphere by the terms of the narrative; this
restriction in verse 8th in fact implying the wider sense of the
word in preceding verses. (2.) Because the atmospheric firmament, elsewhere called heaven, divides the waters above from
those below, whereas it is evident that all these waters, and of
consequence the materials of the atmosphere itself, are included in the earth of the following verse. (3.) Because in verse
14th the sidereal heavens are spoken of as arranged from
pre-existing materials, which refers their actual creation back
to this passage.

In the words now under consideration we therefore regard
the heavens as including the whole material universe beyond
the limits of our earth. That this sense of the word is not
unknown to the writers of Scripture, and that they had enlarged and rational views of the star-spangled abysses of
space, will appear from the terms employed by Moses in his
solemn warning against the Sabæan idolatry, in Deuteronomy
iv. : "And lest thou lift up thine eyes to the heavens, and
when thou seest the sun and the moon and the stars, even all
the host of the heavens, shouldest be incited to worship them
and serve them which Jehovah thy God hath appointed to all

---

* Job xxxviii., 37. † Gen. i., 14 ; Deut. xvii., 3.
‡ Gen. xxviii., 17 ; Job xv., 15 ; Psa. ii., 4.

nations under the whole heavens." To the same effect is the
expression of the awe and wonder of the poet king of Israel
in Psalm viii. :

> " When I consider the heavens, the work of thy fingers,
> The moon and the stars which thou hast ordained;
> What is man that thou art mindful of him?"

I may observe, however, that throughout the Scriptures the
word in question is much more frequently applied to the
atmospheric than to the sidereal heavens. The reason of
this appears in the terms of verse 8th.

If we have correctly referred the term "heavens" to the whole
of extramundane space, then the word "earth" must denote our
globe as a distinct world, with all the liquid and aeriform substances on its surface. The arrangement of the whole universe under the heads "heaven" and "earth" has been derided
as a division into " infinity and an atom ;" but when we consider the relative importance of the earth to us, and that it
constitutes the principal object of the whole revelation to
which this is introductory, the absurdity disappears, and we
recognize the classification as in the circumstances natural
and rational. The word "earth" (*arets*) is, however, generally
used to denote the dry land, or even a region or district of
country. It is indeed expressly restricted to the dry land in
verse 10th; but as in the case of the parallel limitation of the
word "heaven," we may consider this as a hint that its previous
meaning is more extended. That it is really so, appears from
the following considerations : (1.) It includes the deep, or the
material from which the sea and atmosphere were afterwards
formed. (2.) The subsequent verses show that at the period
in question no dry land existed. If instances of a similar
meaning from other parts of Scripture are required, I give

the following : Genesis ii., 1 to 4, " Thus the heavens and the earth were finished, and all the host of them ;" " these are the generations of the heavens and the earth.' In this general summary of the creative work, the earth evidently includes the seas and all that is in them, as well as the dry land ; and the whole expression denotes the universe. The well-known and striking remark of Job, " Who hangeth the earth upon nothing," is also a case in point, and must refer to the whole world, since in other parts of the same book the dry land or continental masses of the earth are said, and with great truth and propriety, to be supported above the waters on pillars or foundations. The following passages may also be cited as instances of the occurrence of the idea of the whole world expressed by the word "earth:" Exodus x., 29, "And Moses said unto him, As soon as I am gone out of the city, I will spread abroad my hands unto the Lord, and the thunder shall cease, neither shall there be any more hail ; that thou mayest know the earth is the Lord's ;" Deuteronomy x., 14, " Behold, the heaven and the heaven of heavens is the Lord's, the earth also, and all that therein is."

The material universe was brought into existence in the "beginning "—a term evidently indefinite as far as regards any known epoch, and implying merely priority to all other recorded events. It can not be the first day, for there is no expressed connection, and the work of the first day is distinct from that of the beginning. It can not be a general term for the whole six days, since these are separated from it by that chaotic or formless state to which we are next introduced. The beginning, therefore, is the threshold of creation—the line that separates the old tenantless condition of space from the world-crowded galaxies of the existing universe. The only other information respecting it that we have in Scrip-

ture is in that fine descriptive poem in Proverbs viii., in
which the Wisdom of God personified—who may be held to
represent the Almighty Word, or Logos, introduced in the
formula "God said," and afterward referred to in Scripture as the manifested or conditioned Deity, the Mediator
between man and the otherwise inaccessible Divinity, the
agent in the work of creation as well as in that of redemption
—narrates the origin of all created things :

> " Jehovah possessed * me, the beginning of his way,
> Before his work of old.
> I was set up from everlasting,
> From the beginning, before the earth was ;
> When there were no deeps I was brought forth,
> When there were no fountains abounding in water."

The beginning here precedes the creation of the earth, as
well as of the deep which encompassed its surface in its
earliest condition.   The beginning, in this point of view,
stretches back from the origin of the world into the depths
of eternity.   It is to us emphatically *the* beginning, because
it witnessed the birth of our material system ; but to the
eternal Jehovah it was but the beginning of a great series of
his operations, and we have no information of its absolute
duration.   From the time when God began to create the
celestial orbs, until that time when it could be said that he
had created the heavens and the earth, countless ages may
have rolled along, and myriads of worlds may have passed
through various stages of existence, and the creation of our
planetary system may have been one of the last acts of that
long beginning.

The author of creation is Elohim, or God in his general

---

* Not " created," as some read.   The verb is *kana*, not *bara.*

aspect to nature and man, and not in that special aspect in reference to the Hebrew commonwealth and to the work of redemption indicated by the name Jehovah (*Iaveh*). We need not enter into the doubtful etymology of the word; but may content ourselves with that supported by many, perhaps the majority of authorities, which gives it the meaning of "Object of dread or adoration," or with that preferred by Gesenius, which makes it mean the "Strong or mighty one." Its plural form has also greatly tried the ingenuity of the commentators. After carefully considering the various hypotheses, such as that of the plural of majesty of the Rabbins, and the primitive polytheism supposed by certain Rationalists, I can see no better reason than an attempt to give a grammatical expression to that plurality in unity indicated by the appearance of the Spirit or breath of God and his Word, or manifested will and power, as distinct agents in the succeeding verses. This was probably always held by the Hebrews in a general form; and was by our Saviour and his apostles specialized in that trinitarian doctrine which enables both John and Paul explicitly to assert the agency of the second person of the Trinity in the creative work.

This elementary trinitarian idea of the first chapter of Genesis may be further stated thus: The name Elohim expresses the absolute unconditioned will and reason—the Godhead. The manifestation of God in creative power, and in the framing and ordering of the cosmos, is represented by the formula "God said"—the equivalent of the Divine Word. The further manifestation of God in love and sympathy with his work is represented by the Breath of God, and by the expression, "God saw that it was good"—operations these of the Divine Spirit.

The aboriginal root of the word Elohim probably lies far

E

back of the Semitic literature, and comes from the natural exclamations "al," "lo," "la," which arise from the spontaneous action of the human vocal organs in the presence of any object of awe or wonder. The plural form may in like manner be simply equivalent to our terms Godhead or Divinity, implying all that is essentially God without specification or distinction of personalities. As Dr. Tayler Lewis well remarks in his "Introduction to Genesis," we should not dismiss such plurals as mere *usus loquendi.* The plural form of the name of God, of the heavens (literally, the "heights"), of the *olamim,* or time-worlds, of the word for life in Genesis (lives), indicates an idea of vastness and diversity not measurable by speech, which must have been impressed on the minds of early men, otherwise these forms would not have arisen. God, heaven, time, life, were to them existences stretching outward to infinity, and not to be denoted by the bare singular form suitable to ordinary objects.

Fairly regarding, then, this ancient form of words, we may hold it as a clear, concise, and accurate enunciation of an ultimate doctrine of the origin of things, which with all our increased knowledge of the history of the earth we are not in a position to replace with any thing better or more probable. On the other hand, this sublime dogma of creation leaves us perfectly free to interrogate nature for ourselves, as to all that it can reveal of the duration and progress of the creative work. But the positive gain which comes from this ancient formula goes far beyond these negative qualities. If received, this one word of the Old Testament is sufficient to deliver us forever from the superstitious dread of nature, and to present it to us as neither self-existent nor omnipotent, but as the mere handiwork of a spiritual Creator to whom we are kin; as not a product of chance or caprice, but as the result of a definite plan of the All-wise; as not a congeries of unconnected facts and processes, but as a cosmos, a well-ordered though complex machine, designed by Him who is the Almighty and the supreme object of reverence. Had this verse alone constituted the whole Bible, this one utterance would, wherever known and received, have been an inestimable boon to mankind; proclaiming deliverance to the captives of every form of nature-worship and idolatry, and fixing that idea of unity of plan in the universe which is the fruitful and stable root of all true progress in science. We owe profound thanks to the old Hebrew prophet for these words—words which have broken from the necks of once superstitious Aryan races chains more galling than those of Egyptian bondage.

CHAPTER V.

THE DESOLATE VOID.

"And the earth was desolate and empty, and darkness was upon the surface of the deep ; and the Spirit of God moved on the surface of the waters."—Genesis i., 2.

WE have here a few bold outlines of a dark and mysterious scene—a condition of the earth of which we have no certain intimation from any other source, except the speculations based on modern discoveries in physical science. It was "unshaped and empty," formless and uninhabited. The words thus translated are sufficiently plain in their meaning. The first is used by Isaiah to denote the desolation of a ruined city, and in Job and the Psalms as characteristic of the wilderness or desert. Both in connection are employed by Isaiah to express the destruction of Idumea, and by Jeremiah in a powerful description of the ruin of nations by God's judgments. When thus united, they form the strongest expression which the Hebrew could supply for solitary, uninhabited desolation, like that of a city reduced to heaps of rubbish, and to the silence and loneliness of utter decay.

In the present connection these words inform us that the earth was in a chaotic state, and unfit for the residence of organized beings. The words themselves suggest the important question : Are they intended to represent this as the original condition of the earth ? Was it a scene of desolation and confusion when it sprang from the hand of its Creator ?

or was this state of ruin consequent on convulsions which
may have been preceded by a very different condition, not
mentioned by the inspired historian? That it may have been
só is rendered possible by the circumstance that the words
employed are generally used to denote the ruin of places
formerly inhabited, and by the want of any necessary connection in time between the first and second verses. It has even
been proposed, though this does violence to the construction,
to read "and the earth became" desolate and empty. Farther, it seems, *à priori*, improbable that the first act of creative power should have resulted in the production of a mere
chaos. The crust of the earth also shows, in its alternations
of strata and organic remains, evidence of a great series of
changes extending over vast periods, and which might, in a
revelation intended for moral purposes, with great propriety
be omitted.

For such reasons some eminent expositors of these words
are disposed to consider the first verse as a title or introduction, and to refer to this period the whole series of geological
changes; and this view has formed one of the most popular
solutions of the apparent discrepancies between the geological
and Scriptural histories of the world. It is evident, however,
that if we continue to view the term "earth" as including the
whole globe, this hypothesis becomes altogether untenable.
The subsequent verses inform us that at the period in question
the earth was covered by a universal ocean, possessed no atmosphere and received no light, and had not entered into its
present relations with the other bodies of our system. No
conceivable convulsions could have effected such changes on
an earth previously possessing these arrangements; and geology assures us that the existing laws and dispositions in
hese respects have prevailed from the earliest periods to

which it can lead us back, and that the modern state of things was not separated from those which preceded it by any such general chaos. To avoid this difficulty, which has been much more strongly felt as these facts have been more and more clearly developed by modern science, it has been held that the word earth may denote only a particular region, temporarily obscured and reduced to ruin, and about to be fitted up, by the operations of the six days, for the residence of man ; and that consequently the narrative of the six days refers not to the original arrangement of the surface, relations, and inhabitants of our planet, but to the retrieval from ruin and repeopling of a limited territory, supposed to have been in Central Asia, and which had been submerged and its atmosphere obscured by aqueous or volcanic vapors. The chief support of this view is the fact, previously noticed, that the word earth is very frequently used in the signification of region, district, country ; to which may be added the supposed necessity for harmonizing the Scriptures with geological discovery, and at the same time viewing the days of creation as literal solar days.

Can we, however, after finding that in verse 1st the term earth must mean the whole world, suddenly restrict it in verse 2d to a limited region. Is it possible that the writer who in verse 10th for the first time intimates a limitation of the meaning of this word, by the solemn announcement, "And God called the *dry land* earth," should in a previous place use it in a much more limited sense without any hint of such restriction. The case stands thus: A writer uses the word earth in the most general sense ; in the next sentence he is supposed, without any intimation of his intention, to use the same word to denote a region or country, and by so doing entirely to change the meaning of his whole discourse from that which would otherwise have attached to it. Yet the same writer when, a few sentences farther on, it becomes necessary for him to use the word earth to denote the dry land as distinguished from the seas, formally and with an assertion of divine authority, intimates the change of meaning. Is not this supposition contrary not only to sound principles of interpretation, but also to common-sense; and would it not tend to render worthless the testimony of a writer to whose diction such inaccuracy must be ascribed. It is in truth to me surprising beyond measure that such a view could ever have obtained currency; and I fear it is to be attributed to a determination, at all hazards and with any amount of violence to the written record, to make geology and religion coincide. Must we then throw aside this simple and convenient method of reconciliation, sanctioned by Chalmers, Smith, Harris, King, Hitchcock, and many other great or respectable names, and on which so many good men complacently rest. Truth obliges us to do so, and to confess that both geology and Scripture refuse to be reconciled on this basis. We may still admit that the lapse of time between the beginning and the first day may have been great; but we must emphatically deny that this interval corresponds with the time indicated by the series of fossiliferous rocks.

Before leaving this part of the subject, I may remark that the desolate and empty condition of the earth was not necessarily a chaotic mass of confusion—*rudis indigestaque moles;* but in reality, when physically considered, may have been a more symmetrical and homogeneous condition than any that it subsequently assumed. If the earth were first a vast globe of vapor, then a liquid spheroid, and then acquired a crust not yet seamed by fissures or broken by corrugations, and eventually covered with a universal ocean, then in each of these early conditions it would, in regard to its form, be a more perfect globe than at any succeeding time. That something of this kind is the intention of our historian is implied in his subsequent statements as to the absence of land and the prevalence of a universal ocean in the immediately succeeding period, which imply that the crust had not yet been ruptured or disturbed, but presented an even and uniform surface, no part of which could project above the comparatively thin fluid envelope.

The second clause introduces a new object—"*the deep.*" Whatever its precise nature, this is evidently something included in the earth of verse 1st, and created with it. The word occurs in other parts of the Hebrew Scriptures in various senses. It often denotes the sea, especially when in an agitated state (Psa. xlii., 8; Job xxxviii., 10). In Psalm cxxxv., however, it is distinguished from the sea: "Whatsoever the Lord pleased, that did he in heaven, in the earth, in the seas, and *in all deeps.*" In other cases it has been supposed to refer to interior recesses of the earth, as when at the deluge "the fountains of the great deep" are said to have been broken up. It is probable, however, that this refers to the ocean. In some places it would appear to mean the atmosphere or its waters; as Prov. viii., 27-29, "When he prepared the heavens, I was there; when he described a circle on the face of the deep, when he established the clouds above, when he strengthened the fountains of the deep." The Septuagint in this passage reads "throne on the winds" and "fountains under the heaven."* Though we can not attach much value to these readings, there seems little reason to doubt that the author of this passage understands by the

---

* The usual Septuagint rendering is *Abyssus.*

deep the atmospheric waters, and not the sea, which he mentions separately. The same meaning must be attached to
the word in another passage of the Book of Proverbs:
"The Lord in wisdom hath founded the earth, by understanding hath he established the heavens; by his knowledge the depths are broken up, and the clouds drop down
the small rain."

In the passage now under consideration, it would seem
that we have both the deep and the waters mentioned, and
this not in a way which would lead us to infer their identity.
The darkness on the surface of the deep and the Spirit of
God on the face of the waters seem to refer to the condition
of two distinct objects at the same time. Neither can the
word here refer to subterranean cavities, for the ascription of
a surface to these, and the statement that they were enveloped
in darkness, would in this case have neither meaning nor use.
For these reasons I am induced to believe that the locality
of the deep or abyss is to be sought, not in the universal
ocean or the interior of the earth, but in the vaporous or aeriform mass mantling the surface of our nascent planet, and
containing the materials out of which the atmosphere was
afterward elaborated. This is a view leading to important
consequences: one of which is that the darkness on the
surface of the deep can not have been, as believed by the
advocates of a local chaos, a mere atmospheric obscuration; since even at the *surface* of what then represented the
atmosphere darkness prevailed. "God covered the earth
with the deep as with a garment, and the waters stood above
the hills," and without this outer garment was the darkness
of space destitute of luminaries, at least of those greater ones
which are of primary importance to us. We learn from the
following verses that there was no layer of clear atmosphere

E 2

in this misty deep, separating the clouds from the ocean waters.

The last clause of the verse has always been obscure, and perhaps it is still impossible to form a clear idea of the operation intended to be described. We are not even certain whether it is intended to represent any thing within the compass of ordinary natural laws, or to denote a direct intervention of the Creator, miraculous in its nature and confined to one period. It is possible that the general intention of the statement may be to the effect that the agency of the divine power in separating the waters from the incumbent vapors had already commenced—that the Spirit which would afterward evoke so many wonders out of the chaotic mass was already acting upon it in an unseen and mysterious way, preparing it for its future destiny.

Some commentators, both Jewish and Christian, are, however, disposed to view the *Ruach Elohim*, Spirit, or breath of God, as meaning a wind of God, or mighty wind, according to a well-known Hebrew idiom. The word in its primary sense means wind or breath, and there are undoubted instances of the expression "wind of God" for a great or strong wind. For example, Isaiah xl., 7 : "The grass withereth because the wind of the Lord bloweth upon it;" see also 2 Kings ii., 16. Such examples, however, are very rare, and by no means sufficient of themselves to establish this interpretation. Those who hold this view do so mainly in consideration of the advantage which it affords in attaching a definite meaning to the expression. Many of them are not, however, aware of its precise import in a cosmical point of view. A violent wind, before the formation of the atmosphere, and the establishment of the laws which regulate the suspension and motions of aqueous vapors and clouds, must have been merely

an agitation of the confused misty and vaporous mass of the
deep ; since, as Ainsworth—more careful than modern interpreters—long ago observed, "winde (which is the moving of
the aier) was not created till the second day, that the firmament was spred, and the aier made." Such an agitation is by
no means improbable. It would be a very likely accompani-
ment of a boiling ocean, resting on a heated surface, and of
excessive condensation of moisture in the upper regions of the
atmosphere ; and might act as an influential means of preparing the earth for the operations of the second day. It is
curious also that the Phœnician cosmogony is said to have
contained the idea of a mighty wind in connection with this
part of creation, and the idea of seething or commotion in the
primitive chaos also occurs in the Assyrian tablets of creation, while the Quiché legend represents Hurakon, the storm-
god, as specially concerned in the creative work.* On the
other hand, the verb used in the text rather expresses hovering or brooding than violent motion, and this better corresponds with the old fable of the mundane egg, which seems
to have been derived from the event recorded in this verse.
The more evangelical view, which supposes the Holy Spirit
to be intended, is also more in accordance with the general
scope of the Scripture teachings on this subject; and the
opposite idea is, as Calvin well says, "too frigid" to meet with
much favor from evangelical theologians.

Chaos, the equivalent of the Hebrew "desolation and
emptiness," figures largely in all ancient cosmogonies. That
of the Egyptians is interesting, not only from its resemblance
to the Hebrew doctrine, but also from its probable connec-

* Smith, "Assyrian Genesis." Brasseur de Bourbourg's translation of
the " Popol Vuh" of the ancient Central American Indians.

tion with the cosmogony of the Greeks. Taking the version
of Diodorus Siculus, which though comparatively modern, yet
corresponds with the hints derived from older sources, we
find the original chaos to have been an intermingled condition of elements constituting heaven and earth. This is the
Hebrew "deep." The first step of progress is the separation
of these ; the fiery particles ascending above, and not only
producing light, but the revolution of the heavenly bodies—a
curious foreshadowing of the nebular hypothesis of modern
astronomy. After these, in the terms of the lines quoted by
Diodorus from Euripides, plants, birds, mammals, and finally
man are produced, not however by a direct creative fiat, but
by the spontaneous fecundity of the teeming earth. The
Phœnician cosmogony attributed to Sancuniathon has the
void, the deep, and the brooding Spirit; and one of the terms
employed, "baau," is the same with the Hebrew "bohu,"
void, if read without the points. The Babylonians, according
to Berosus, believed in a chaos—which, however, like the lit-
eral-day theory of some moderns, produced many monsters before Belus intervened to separate heaven and earth. But the
Assyrian legend found in the Ninevch tablets is very precise
in its intimation of the Chaos or *Tiamat*, the mother of all
things ; and, farther, it recognizes this personified chaos as
the principle of evil, whose " dragon " becomes the tempter
of the progenitors of mankind, exactly like the Biblical serpent. This " dragon of the abyss " is thus identical in name
and function with the evil principle even of the last book of the
New Testament, and we have in this also probably the origin
of the Ahriman of the Avesta. Thus in these Eastern theologies the primeval chaos becomes the type of evil as opposed
to the order, beauty, and goodness of the creation of God—a
very natural association ; but one kept in the background by

the Hebrew Scriptures, as tending to a dualistic belief subversive of monotheism. The Greek myth of Chaos, and its children Erebus and Night, who give birth to Aether and Day, is the same tradition, personified after the fanciful manner of a people who, in the primitive period of their civilization, had no profound appreciation of nature, but were full of human sympathies.* Lastly, in a hymn translated by Dr.

---

* It is impossible to avoid recognizing in the Greek Theogony, as it appears in Hesiod and the Orphic poems, an inextricable intermingling of a cosmogony akin to that of Moses with legendary stories of deceased ancestors ; and this has, I must confess, always appeared to me to be a more rational way of accounting for it than its reference to mere nature-myths. Chaos, or space, for the chaos of Hesiod differs from that of Ovid, came first, then Gaea, the earth, and Tartarus, or the lower world. Chaos gave birth to Erebos (identical with the Hebrew Ereb or Erev, evening) and Nyx, or night. These again give birth to Aether, the equivalent of the Hebrew expanse or firmament, and to Hemera, the day, and then the heavenly bodies were perfected. So far the legend is apparently based on some primitive history of creation, not essentially different from that of the Bible. But the Greek Theogony here skips suddenly to the human period ; and under the fables of the marriage of Gaea and Uranos, and the Titans, appears to present to us the antediluvian world, with its intermarriages of the sons of God and men, and its Nephelim or Giants, with their mechanic arts and their crimes. Beyond this, in Kronos and his three sons, and in the strange history of Zeus, the chief of these, we have a coarse and fanciful version of the story of the family of Noah, the insult offered by Ham to his father, and the subsequent quarrels and dispersion of mankind. The Zeus of Homer appears to be the elder of the three, or Japheth, the real father of the Greeks, according to the Bible ; but in the time of Hesiod Zeus was the youngest, perhaps indicating that the worship of the Egyptian Zeus, Ammon or Ham, had already supplanted among the Greeks that of their own ancestor. But it is curious that even in the Bible, though Japhet is said to be the greater, he is placed last in the lists. After the introduction of Greek savans and literati to Egypt, about B.C. 660, they began to regard their own mythology from this point of view, though obliged to be reserved on the subject. The cosmology of Thales, the astronomy of Anaxagoras, and the history of Herodotus afford early evidence of this, and it abounds in later writers. I may refer the reader to Grote (History of Greece, vol. i.) for an able and agreeable sum-

Max Müller from the Rig-Veda, a work probably far older
than the Institutes of Menu, we have such utterances as the
following :

> "Nor aught nor nought existed : yon bright sky
> Was not, nor heaven's broad woof outstretched above.
> What covered all ? what sheltered ? what concealed ?
> Was it the water's fathomless abyss ? * * *
> Darkness there was, and all at first was veiled
> In gloom profound—an ocean without light ;
> The germ that still lay covered in the husk
> Burst forth, one nature, from the fervent heat."

It is evident that the state of our planet which we have just
been considering is one of which we can scarcely form any
adequate conception, and science can in no way aid us, except by suggesting hypotheses or conjectures.   It is remarkable, however, that nearly all the cosmological theories which
have been devised contain some of the elements of the inspired narrative.   The words of Moses appear to suggest a
heated and cooling globe, its crust as yet unbroken by internal forces, covered by a universal ocean, on which rested a
mass of confused vaporous substances ; and it is of such materials, thus combined by the sacred historian, that cosmolo-
gists have built up their several theories, aqueous or igneous,
of the early state of the earth.   Geology, as a science of ob-

---

mary of this subject ; and may add that even the few coincidences above
pointed out between Greek mythology and the Bible, independently of the
multitudes of more doubtful character to be found in the older writers on
this subject, appear very wonderful, when we consider that among the
Greeks these vestiges of primitive religion, whether brought with them
from the East or received from abroad, must have been handed down for
a long time by oral tradition among the people ; but obscure though they
may be, the circumstance that some old writers have ridden the resemblances to death affords no excuse for the prevailing neglect of them
in more modern times.

servation and induction, does not carry us back to this period.
It must still and always say, with Hutton, that it can find " no
trace of a beginning, no prospect of an end "—not because
there has been no beginning or will be no end, but because
the facts which it collects extend neither to the one nor the
other. Geology, like every other department of natural history, can but investigate the facts which are open to observation, and reason on these in accordance with the known laws
and arrangements of existing nature. It finds these laws to
hold for the oldest period to which the rocky archives of the
earth extend. Respecting the origin of these general laws
and arrangements, or the condition of the earth before they
originated, it knows nothing. In like manner a botanist
may determine the age of a forest by counting the growth
rings of the oldest trees, but he can tell nothing of the forests
that may have preceded it, or of the condition of the surface
before it supported a forest. So the archæologist may on
Egyptian monuments read the names and history of successive dynasties of kings, but he can tell nothing of the state
of the country and its native tribes before those dynasties
began or their monuments were built. Yet geology at least
establishes a probability that a time was when organized beings did not exist, and when many of the arrangements of the
surface of our earth had not been perfected ; and the few
facts which have given birth to the theories promulgated on
this subject tend to show that this pre-geological condition
of the earth may have been such as that described in the
words now under consideration. I may remark, in addition, that if the words of Moses imply the cooling of the
globe from a molten or intensely heated state down to a
temperature at which water could exist on its surface, the
known rate of cooling of bodies of the dimensions and mate-

rials of the earth shows that the time included in these two verses of Genesis must have been enormous, amounting it may be to many millions of years.

There are two other sciences besides geology which have in modern times attempted to penetrate into the mysteries of the primitive abyss, at least by hypothetical explanations—astronomy and chemistry. The magnificent nebular hypothesis of La Place, which explains the formation of the whole solar system by the condensation of a revolving mass of gaseous matter, would manifestly bring our earth to the condition of a fluid body, with or without a solid crust, and surrounded by a huge atmosphere of its more volatile materials, gradually condensing itself around the central nucleus. Chemistry informs us that this vaporous mass would contain not only the atmospheric air and water, but all the carbon, sulphur, phosphorus, chlorine, and other elements, volatile in themselves, or forming volatile compounds with oxygen or hydrogen, that are now imprisoned in various states of combination in the solid crust of the earth. Such an atmosphere—vast, dark, pestilential, and capable in its condensation of producing the most intense chemical action—is a necessity of an earth condensing from a vaporous and incandescent state. Thus, in so far as scientific speculation ventures to penetrate into the genesis of the earth, its conclusions are at one with the Mosaic cosmogony and with the traditions of most ancient nations as to the primitive existence of a chaos—formless and void, in which "nor aught nor nought existed."

Some of the details of the Mosaic vision of the primeval chaos may be supplied by the probabilities established by physics and chemistry. Our first idea of the earth would be a vast vaporous ball, recently spun out from the general mass of vapors forming the nebula which once represented the solar

system. This huge cloud, whirling its annual round about
the still vaporous centre of the system, would consist of all
the materials now constituting the solid rocks as well as those
of the seas and atmosphere, their atoms kept asunder by the
force of heat, preventing not only their mechanical union, but
even their chemical combination. But heat is being radiated
on all sides into space, and the opposing force of gravitation
is little by little gathering the particles toward the centre. At
length a liquid nucleus is formed, while upon this are being
precipitated showers of condensing matter from the still vast
atmosphere to add to its volume. As this process advances,
a new brilliancy is given to the feebly shining vapors by the
incandescence of solid particles in the upper layers of the
atmosphere, and in this stage our earth would be a little sun, a
miniature of that which now forms the centre of our system,
and which still, by virtue of its greater mass, continues in this
state. But at length, by further cooling, this brilliancy is lost,
and the still fluid globe is surrounded by a vast cloudy pall,
in which condensing vapors gather in huge dark masses, and
amid terrible electrical explosions, pour, in constantly increasing, acid, corrosive rains, upon the heated nucleus, combining
with its materials, or again flashing into vapors. Thus darkness dense and gross would settle upon the vaporous deep,
and would continue for long ages, until the atmosphere could
be finally cleared of its superfluous vapors. In the mean
time a crust of slag or cinder has been forming upon the
molten nucleus. Broken again and again by the heaving of
the seething mass, it at length sets permanently, and finally
allows some portion of the liquid rain condensed upon it to
remain as a boiling ocean. Then began the reign of the
waters, under which the first stratified rocks were laid down
by the deposit of earthy and saline matter suspended or dis-

solved in the heated sea. Such is the picture which science presents to us of the genesis of the earth, and so far as we can judge from his words, such must have been the picture presented to the mental vision of the ancient seer of creation; but he could discern also that mysterious influence, the "breath of Elohim," which moved on the face of the waters, and prepared for the evolution of land and of life from their bosom. He saw—

> "An earth—formless and void;
> A vaporous abyss—dark at its very surface;
> A universal ocean—the breath of God hovering over it."

How could such a scene be represented in words? since it presented none of the familiar features of the actual world. Had he attempted to dilate upon it, he would, in the absence of the facts furnished by modern science, have been obliged, like the writers of some of the less simple and primitive cosmogonies already quoted,* to adopt the feeble expedient of enumerating the things not present. He wisely contents himself with a few well-chosen words, which boldly sketch the crude materials of a world hopeless and chaotic but for the animating breath of the Almighty, who has created even that old chaos out of which is to be worked in the course of the six creative days all the variety and beauty of a finished world.

In conclusion, the reader will perceive how this reticence of the author of Genesis strengthens the argument for the primitive age of the document, and for the vision-theory as to its origin; and will also observe that, in the conception of this ancient writer, the "promise and potency" of order and life reside not alone in the atoms of a vaporous world, but also in the will of its Creator.

---

* Pages 21, 22, and 109, *supra.*

CHAPTER VI.

LIGHT AND CREATIVE DAYS.

"And God said, Let light be, and light was ; and God saw the light that it was good, and separated the light from the darkness : and God called the light Day ; and the darkness he called Night. And Evening was and Morning was—Day one."—Genesis i., 3-5.

LIGHT is the first element of order and perfection introduced upon our planet—the first innovation on the old régime of darkness and desolation. There is a beautiful propriety in this, for the Hebrew *Aur* (light) should be viewed as including heat and electricity as well as light ; and these three forces—if they are really distinct, and not merely various movements of one and the same ether—are in themselves, or the proximate causes of their manifestation, the prime movers of the machinery of nature, the vivifying forces without which the primeval desolation would have been eternal. The statement presented here is, however, a bold one. Light without luminaries, which were afterward formed—independent light, so to speak, shining all around the earth—is an idea not likely to have occurred in the days of Moses to the framer of a fictitious cosmogony, and yet it corresponds in a remarkable manner with some of the theories which have grown out of modern induction.

I have said that the Hebrew word translated "light" includes the vibratory movements which we call heat and electricity as well. I make this statement, not intending to assert that the Hebrews experimented on these forces in the manner of modern science, and would therefore be prepared to understand their laws or correlations as fully as we can. I give the word this general sense simply because throughout the Bible it is used to denote the solar light and heat, and also the electric light of the thunder-cloud : "the light of His cloud," "the bright light which is in the clouds." The absence of "*aur*," therefore, in the primeval earth, is the absence of solar radiation, of the lightning's flash, and of volcanic fires. We shall in the succeeding verses find additional reasons for excluding all these phenomena from the darkness of the primeval night.

The light of the first day can not reasonably be supposed to have been in any other than a visible and active state. Whether light be, as supposed by the older physicists, luminous matter radiated with immense velocity, or, as now appears more probable, merely the undulations of a universally diffused ether, its motion had already commenced. The idea of the matter of light as distinct from its power of affecting the senses does not appear in the Scriptures any farther than that the Hebrew name is probably radically identical with the word ether now used to express the undulating medium by which light is propagated ; and if it did, the general creation of matter being stated in verse 1, and the notice of the separation of light and darkness being distinctly given in the present verse, there is no place left for such a view here. For this reason, that explanation of these words which supposes that on the first day the *matter* of light, or the ether whose motions produce light, was created, and that on the fourth day, when luminaries were appointed, it became visible by beginning to undulate, must be abandoned ; and the connection between these two statements must be sought in some other group of facts than that connected with the existence of the matter of light as distinct from its undulations.

What, then, was the nature of the light which on the first day shone without the presence of any local luminary? It must have proceeded from luminous matter diffused through the whole space of the solar system, or surrounding our globe as with a mantle. It was "clothed with light as with a garment,"

"Sphered in a radiant cloud, for yet the sun was not."

We have already rejected the hypothesis that the primeval night proceeded from a temporary obscuration of the atmosphere; and the expression, "God said, Let light be," affords an additional reason, since, in accordance with the strict precision of language which everywhere prevails in this ancient document, a mere restoration of light would not be stated in such terms. If we wish to find a natural explanation of the mode of illumination referred to, we must recur to one or other of the suppositions mentioned above, that the luminous matter formed a nebulous atmosphere, slowly concentrating itself toward the centre of the solar system, or that it formed a special envelope of our earth, which subsequently disappeared.

We may suppose this light-giving matter to be the same with that which now surrounds the sun, and constitutes the stratum of luminous substance which, by its wondrous and unceasing power of emitting light, gives him all his glory. To explain the division of the light from the darkness, we need only suppose that the luminous matter, in the progress of its concentration, was at length all gathered within the earth's orbit, and then, as one hemisphere only would be illuminated at a time, the separation of light from darkness, or of day from night, would be established. This hypothesis, suggested by the words themselves, affords a simple and natural explanation of a statement otherwise obscure.

It is an instructive circumstance that the probabilities respecting the early state of our planet, thus deduced from the Scriptural narrative, correspond very closely with the most ingenious and truly philosophical speculation ever hazarded respecting the origin of our solar system.   I refer to the cosmical hypothesis of La Place, which was certainly formed without any reference to the Bible ; and by persons whose views of the Mosaic narrative are of that shallow character which is too prevalent, has been suspected as of infidel tendency.   La Place's theory is based on the following properties of the solar system, which will be found referred to in this connection in many popular works on astronomy : 1. The orbits of the planets are nearly circular. 2. They revolve nearly in the plane of the sun's equator.* 3. They all revolve round the sun in one direction, which is also the direction of the sun's rotation.   4. They rotate on their axes also, as far as is known, in the same direction. 5. Their satellites, with the exception of those of Uranus and Neptune, revolve in the same direction.   Now all these coincidences can scarcely have been fortuitous, and yet they might have been otherwise without affecting the working of the system ; and, farther, if not fortuitous, they correspond

---

* The minor planets discovered in more recent times between Mars and Jupiter form an exception to this; but they are of little importance, and exceptional in other respects as well.   To give their arrangement and the motions of the satellites of Uranus, would require the further assumption of some unknown disturbing cause.

precisely with the results which would flow from the condensation of a revolving mass of nebulous matter. La Place,
therefore, conceived that in the beginning the matter of our
system existed in the condition of a mass of vaporous material, having a central nucleus more or less dense, and the
whole rotating in a uniform direction. Such a mass must,
"in condensing by cold, leave in the plane of its equator
zones of vapor composed of substances which required an
intense degree of cold to return to a liquid or solid state.
These zones must have begun by circulating round the sun
in the form of concentric rings, the most volatile molecules
of which must have formed the superior part, and the most
condensed the inferior part. If all the nebulous molecules
of which these rings are composed had continued to cool
without disuniting, they would have ended by forming a
liquid or solid ring. But the regular constitution which all
parts of the ring would require for this, and which they would
have needed to preserve when cooling, would make this phenomenon extremely rare. Accordingly the solar system presents only one instance of it—that of the rings of Saturn.
Generally the ring must have broken into several parts which
have continued to circulate round the sun, and with almost
equal velocity, while at the same time, in consequence of
their separation, they would acquire a rotatory motion round
their respective centres of gravity; and as the molecules of
the superior part of the ring—that is to say, those farthest
from the centre of the sun—had necessarily an absolute
velocity greater than the molecules of the inferior part which
is nearest it, the rotatory motion common to all the fragments
must always have been in the same direction with the orbit-
ual motion. However, if after their division one of these
fragments has been sufficiently superior to the others to unite

them to it by its attraction, they will have formed only a mass
of vapor, which, by the continual friction of all its parts,
must have assumed the form of a spheroid, flattened at the
poles and expanded in the direction of its equator."*   Here,
then, are rings of vapor left by the successive retreats of the
atmosphere of the sun, changed into so many planets in the
condition of vapor, circulating round the central orb, and
possessing a rotatory motion in the direction of their revolution, while the solar mass was gradually contracting itself
round its centre and assuming its present organized form.
Such is a general view of the hypothesis of La Place, which
may also be followed out into all the known details of the
solar system, and will be found to account for them all.
Into these details, however, we can not now enter.   Let us
now compare this ingenious speculation with the Scripture
narrative.   In both we have the raw material of the heavens
and the earth created before it assumed its distinct forms.
In both we have that state of the planets characterized as
without form and void, the condensing nebulous mass of
La Place's theory being in perfect correspondence with the
Scriptural "deep."   In both it is implied that the permanent
mutual relations of the several bodies of the system must
have been perfected long after their origin.   Lastly, supposing the luminous atmosphere of our sun to have been of such
a character as to concentrate itself wholly around the centre
of the system, and that as it became concentrated it acquired
its intense luminosity, we have in both the production of light
from the same cause ; and in both it would follow that the
concentration of this matter within the orbit of the earth
would effect the separation of day from night, by illuminating

---

* Nichol's " Planetary System. "

alternately the opposite sides of the earth. It is true that the theory of La Place does not provide for any such special condensation of luminous matter, nor for any precise stage of the process as that in which the arrangements of light and darkness should be completed ; but under his hypothesis it seems necessary to account in some such way for the sole luminosity of the sun ; and the point of separation of day and night must have been a marked epoch in the history of the process for each planet. The theory of accretion of matter which has in modern times been associated with that of La Place would equally well accord with the indications in our Mosaic record.*

It is further to be observed that so long as the material of the earth constituted a part of the great vaporous mass, it would be encompassed with its diffused light, and that after it had been left outside the contracting solar envelope, it might still retain some independent luminosity in its atmosphere, a trace of which may still exist in the auroral displays of the upper strata of the air. The earth might thus at first be in total darkness. It might then be dimly lighted by the surrounding nebulosity, or by a luminous envelope in its own atmosphere. Then it might, as before explained, relapse into the darkness of its misty mantle, and as this cleared away and the light of the sun increased and became condensed, the latter would gradually be installed into his office as the sole orb of day. It is quite evident that we thus have a sufficient hypothetical explanation of the light of the first of the creative æons ; and this is all that in the present state of science we can expect. "Where is the way where light dwelleth? and as for darkness, where is the place thereof, that thou shouldest

---

* Proctor's Lectures, etc.

F

take it to the bound thereof, and know the way to the house
thereof?"

For the reasons above given, we must regard the hypothesis of the great French astronomer as a wonderful approximation to the grand and simple plan of the construction of
our system as revealed in Scripture.   Nor must we omit to
notice that the telescope and the spectroscope reveal to us in
the heavens gaseous nebular bodies which may well be new
systems in progress of formation, and in which the Creator is
even now dividing the light from the darkness.   Still another
thought in connection with this subject is that the theory of
a condensing system affords a measure of the aggregate time
occupied in the work of creation.   Sir William Thomson's
well-known calculations give us one hundred millions of
years as the possible age of the earth as a planetary globe;
but calculations of the sun's heat as produced by gravitation
alone would give a much less time.   We have, however, a
right to assume an original heated condition of the vaporous
mass from which the sun was formed.   Still the date above
given would seem to be a maximum rather than a minimum
age for the solar system.

"God saw the light that it was good," though it illuminated
but a waste of lifeless waters.   It was good because beautiful
in itself, and because God saw it in its relations to long trains
of processes and wonderful organic structures on which it was
to act as a vivifying agency.   Throughout the Scriptures light
is not only good, but an emblem of higher good.   In Psalm
civ. God is represented as " clothing himself with light as with
a garment ;" and in many other parts of these exquisite lyrics
we have similar figures.   "The Lord is my light and salvation ;" " Lift up the light of thy countenance upon me ;"
" The entrance of thy law giveth light ;" " The path of the

just is as a shining light." And the great spiritual Light of the world, the "only begotten of the Father," the mediator alike in creation and redemption, is himself the "Sun of Righteousness." Perhaps the noblest Scripture passage relating to the blessing of light is one in the address of Jehovah to Job, which is unfortunately so imperfectly translated in the English version as to be almost unintelligible :

> "Hast thou in thy lifetime given law to the morning,
> Or caused the dawn to know its place,
> That it may enclose the horizon in its grasp,
> And chase the robbers before it :
> It rolls along as the seal over the clay,
> Causing all things to stand forth in gorgeous apparel."*
>
> Job xxxviii., 12.

The concluding words, "Day one," bring us to the consideration of one of the most difficult problems in this history, and one on which its significance in a great measure depends—the meaning of the word *day*, and the length of the days of creation.

In pursuing this investigation, I shall refrain from noticing in detail the views of the many able modern writers who, from Cuvier, De Luc, and Jameson, down to Hugh Miller, Donald McDonald, and Tayler Lewis, have maintained the period theory, or those equally numerous and able writers who have supported the opposite view. I acknowledge obligations to them all, but prefer to direct my attention immediately to the record itself.

The first important fact that strikes us is one which has

---

* This translation is as literal as is consistent with the bold abruptness of the original. The last idea is that of a cylindrical seal rolling over clay, and leaving behind a beautiful impression where all before was a blank.

not received the attention it deserves, viz., that the word *day* is evidently used in three senses in the record itself. We are told (verse 5th) that God called the *light*, that is, the diurnal continuance of light, day. We are also informed that the *evening* and the *morning* were the first day. Day, therefore, in one of these clauses is the light as separated from the darkness, which we may call the *natural day;* in the other it is the whole time occupied in the creation of light and its separation from the darkness, whether that was a *civil or astronomical day* of twenty-four hours or some longer period. In other words, the daylight, to which God is represented as restricting the use of the term day, is only a part of a day of creation, which included both light and darkness, and which might be either a civil day or a longer period, but could not be the natural day intervening between sunrise and sunset, which is the *ordinary* day of Scripture phraseology. Again, in the 4th verse of chapter ii., which begins the second part of the history, the whole creative week is called one day— "In the day that Jehovah Elohim made the earth and the heavens." Such an expression must surely in such a place imply more than a mere inadvertence on the part of the writer or writers.

To pave the way for a right understanding of the day of creation, it may be well to consider, in the first place, the manner in which the *shorter day* is introduced. In the expression, "God *called* the light day," we find for the first time the Creator naming his works, and we may infer that some important purpose was to be served by this. The nature of this purpose we ascertain by comparison with other instances of the same kind occurring in the chapter. God called the darkness night, the firmament heaven, the dry land earth, the gathered waters seas. In all these cases the purpose seems to have been one of verbal definition, perhaps along with an assertion of sovereignty. It was necessary to distinguish the diurnal darkness from that unvaried darkness which had been of old, and to discriminate between the limited waters of an earth having dry land on its surface and those of the ancient universal ocean. This is effected by introducing two new terms, night and seas. In like manner it was necessary to mark the new application of the term earth to the dry land, and that of heaven to the atmosphere, more especially as these were the senses in which the words were to be popularly used. The intention, therefore, in all these cases was to affix to certain things names different from those which they had previously borne in the narrative, and to certain terms new senses differing from those in which they had been previously used. Applying this explanation here, it results that the probable reason for calling the light day is to point out that the word occurs in two senses, and that while it was to be the popular and proper term for the natural day, this sense must be distinguished from its other meaning as a day of creation. In short, we may take this as a plain and authoritative declaration *that the day of creation is not the day of popular speech.* We see in this a striking instance of the general truth that in the simplicity of the structure of this record we find not carelessness, but studied and severe precision, and are warned against the neglect of the smallest peculiarities in its diction.

What, then, is the day of creation, as distinguished by Moses himself from the natural day. The general opinion, and that which at first sight appears most probable, is that it is merely the ordinary civil day of twenty-four hours. Those who adopt this view insist on the impropriety of diverting the word from its usual sense. Unfortunately, however, for this argument, the word is not very frequently used in the Scriptures for the whole twenty-four hours of the earth's revolution. Its etymology gives it the sense of the time of glowing or warmth, and in accordance with this the divine authority here limits its meaning to the daylight. Accordingly throughout the Hebrew Scriptures *yom* is generally the natural and not the civil day ; and where the latter is intended, the compound terms "day and night" and "evening and morning" are frequently used. Any one who glances over the word "day" in a good English concordance can satisfy himself of this fact. But the sense of natural day from sunrise to sunset is expressly excluded here by the context, as already shown ; and all that we can say in favor of the interpretation that limits the day of creation to twenty-four hours, is that next to the use of the word for the natural day, which is its true popular meaning, its use for the civil day is perhaps the most frequent. It is therefore by no means a statement of the whole truth to affirm, as many writers have done, that the civil day is *the ordinary* meaning of the term. At the same time we may admit that this is *one* of its ordinary meanings, and therefore may be its meaning here. Another argument frequently urged is that the day of creation is said to have had an evening and morning. We shall consider this more fully in the sequel, and in the mean time may observe that it appears rather hazardous to attribute an ordinary evening and morning to a day which, on the face of the record, preceded the formation and arrangement of the luminaries which are " for days and for years."*

---

* Professor Dana thus sums up the various meanings of the word *day* in Genesis : " *First*, in verse 5, the *light* in general is called day, the darkness night. *Second*, in the same verse, *evening and morning* make the first day, before the sun appears. *Third*, in verse 14, day stands for *twelve hours*,

But it may be affirmed that in the Bible long and undefined periods are indicated by the word "day." In many of
these cases the word is in the plural : as Genesis iv., 3, " And
after days it came to pass," rendered in our version " in process of time ;" Genesis xl., 4, " days in ward," rendered " a season." Such instances as these are not applicable to the present question, since the plural may have the sense of indefinite
time, merely by denoting an undetermined number of natural
days. Passages in which the singular occurs in this sense
are those which strictly apply to the case in hand, and such
are by no means rare. A very remarkable example is that
in Genesis ii., 4, already mentioned, where we find, " In the
day when Jehovah Elohim made the earth and the heavens."
This day must either mean the beginning, or must include
the whole six days; most probably the latter, since the word
" made " refers not to the act of creation, properly so called,
but to the elaborating processes of the creative week ; and
occurring as this does immediately after the narrative of creation, it seems almost like an intentional intimation of the
wide import of the creative days. It has been objected, however, that the expression " in the day " is properly a compound adverb, having the force of " when " or " at the time."
But the learned and ingenious authors who urge this objection have omitted to consider the relative probabilities as to

---

or the period of daylight, as dependent on the sun. *Fourth*, same verse,
in the phrase " days and seasons," day stands for a period of *twenty-four*
hours. *Fifth*, at the close of the account, in verse 4 of the second chapter,
day means the *whole period of creation*. These uses are the same that we
have in our own language.

Warring, in his book " The Miracle of To-day," has suggested that the
Mosaic days are *epochal* days, each considered as the close and culmina-
tion of a period. This is an ingenious suggestion, and very well coincides
with the day-period theory as defended in the text.

whether the adverbial use had arisen while the word *yom* meant simply a day, or whether the use of the noun for long periods was the reason of the introduction of such an adverbial expression. The probabilities are in favor of the latter, for it is not likely that men would construct an adverb referring to indefinite time from a word denoting one of the most precisely limited portions of time, unless that word had also a second and more unlimited sense. Admitting, therefore, that the phrase is an adverb of time, its use so early as the date of the composition of Genesis, to denote a period longer than a literal day, seems to imply that this indefinite use of the word was of high antiquity, and probably preceded the invention of any term by which long periods could be denoted.

This use of the word "day" is, however, not limited to cases of the occurrence of the formula "in the day."   The following are a few out of many instances that might be quoted: Job xviii., 20, "They that come after him shall be astonished at his day;" Job xv., 32, "It shall be accomplished before his *time;*" Judges xviii., 30, "Until the day of the captivity of the land;" Deut. i., 39, "And your children which in that day had no knowledge of good and evil;" Gen. xxxix., 10, "And it came to pass about that time" (on that day). We find also abundance of such expressions as "day of calamity," "day of distress," "day of wrath," "day of God's power," "day of prosperity."   In such passages the word is evidently used in the sense of era or period of time, and this in prose as well as poetry.

There is a remarkable passage in the Psalms, which conveys the idea of a day of God as distinct from human or terrestrial days :

> "Before the mountains were brought forth,
> Or ever thou hadst formed the earth and the world,

Even from everlasting to everlasting, thou art God.
Thou turnest man to destruction,
And sayest, Return, ye children of men ;
For a thousand years are in thy sight as yesterday when it is past,
And as a watch in the night."*

It is a singular coincidence that the authorship of this
Psalm is attributed to Moses, and that its style and language
correspond with the songs credited to him in Deuteronomy.
It is farther to be observed that the reference is to the long
periods employed in creation as contrasted with the limited
space of years allotted to man. Its meaning, too, is somewhat obscured by the inaccurate translation of the third line.
In the original it is, "From *olam* to *olam* thou art, O El"—that
is, "from age to age." These long ages of creation, constituting a duration to us relatively eternal, were so protracted
that even a thousand years are but as a watch in the night.
If this Psalm is rightly attributed to the author of the first
chapter of Genesis, it seems absolutely certain that he understood his own creative days as being *Olamim* or æons. The
same thought occurs in the Second Epistle of Peter: "One
day is with the Lord as a thousand years, and a thousand
years as one day."

That the other writers of the Old Testament understood
the creative days in this sense, might be inferred from the entire absence of any reference to the work of creation as short,
since it occupied only six days. Such reference we may find
in modern writers, but never in the Scriptures. On the contrary, we receive the impression of the creative work as long
continued. Thus the divine Wisdom says in Prov. viii., The
Lord possessed me "from the beginning of his way before

---

* Psalm xc.

F 2

his works of old, from everlasting, before the antiquities of the earth." So in Psalm cxlv., God's kingdom relatively to nature and providence is a kingdom "of all ages." In Psalm civ., which is a poetical version of the creative work, and the oldest extant commentary on Genesis i., it is evident that there was no idea in the mind of the writer of a short time, but rather of long consecutive processes; and I may remark here that the course of the narrative itself in Genesis i., implies time for the replenishing of the earth with various forms of being in preparation for others, exactly as in Psalm civ.

Perhaps one of the most conclusive arguments in favor of the length of the creative days is that furnished by the seventh day and the institution of the Sabbath. In Genesis the seventh day is not said to have had an evening or morning, nor is God said to have resumed his work on any eighth day. Consequently the seventh day of creation must be still current. Now in the fourth commandment the Israelites are enjoined to "remember the Sabbath-day," because "in six days God created the heavens and the earth." Observe here that the Sabbath is to be remembered as an institution already known. Observe farther that the commandment is placed in the middle of the Decalogue, a solitary piece of apparently arbitrary ritual amid the plainest and most obvious moral duties. Observe also that the reason given— namely, God's six days' work and seventh day's rest—seems at first sight both far-fetched and trivial, as an argument for abstaining from work in a seventh part of our time. How is all this to be explained? Simply, I think, on the supposition that the Lawgiver, and those for whom he legislated, knew beforehand the history of creation and the fall, as we have them recorded in Genesis, and knew that God's days are æons. The argument is not, "God worked on six natural days, and

rested on the seventh; do you therefore the same." Such
an argument could have no moral or religious force, more
especially as it could not be affirmed that God habitually
works and rests in this way. The argument reaches far deeper and higher. It is this. God created the world in six of
his days, and on the seventh rested, and invited man in Eden
to enter on his rest as a perpetual Sabbath of happiness.
But man fell, and lost God's Sabbath. Therefore a weekly
Sabbath was prescribed to him as a memorial of what he had
lost, and a pledge of what God has promised in the renewal
of life and happiness through our Saviour. Thus the Sabbath
is the central point of the moral law—the Gospel in the Decalogue—the connection between God and man through the
promise of redemption. It is this and this alone that gives it
its true religious significance, but is lost on the natural-day
theory. It would farther seem that this view of the law was
that of our Lord himself, and was known to the Jews of his
time, for, when blamed for healing a man on the Sabbath, he
says, "My Father worketh hitherto, and I work"—an argument whose force depended on the fact that God continues to
work in his providence throughout his long Sabbath, which
has never been broken except by man. Farther, the writer
of the Epistle to the Hebrews takes this view in arguing as to
the rest or Sabbatism that remains to the people of God. His
argument (chap. iv., 4) may be stated thus: God finished his
work and entered into his rest. Man, in consequence of the
fall, failed to do so. He has made several attempts since, but
unsuccessfully. Now Christ has finished his work, and has
entered into his Sabbath, and through him we may enter into
that rest of God which otherwise we can not attain to. This
does not, it is true, refer to the keeping of a Sabbath-day;
but it implies an understanding of the reference to God's

olamic Sabbath, and also implies that Christ, having entered
into his Sabbatism in heaven, gives us a warrant for the Christian Sabbath or Lord's day, which has the same relation to
Christ's present Sabbatism in heaven that the old Sabbath
had to God's rest from his work of creation. *

We may add to these considerations the use of the Greek
term *Aïon* in the New Testament, for what may be called
time-worlds as distinguished from space-worlds.   For example, take the expression in Heb. i., 2: "His Son, by whom
he made the worlds," or, literally, "constituted the æons"—
the long time-worlds of the creation.   For God's worlds must
exist in time as well as in space, and both may to our minds
alike appear as infinities.   If, then, we find that Moses himself
seems to have understood his creative days as æons, that the
succeeding Old Testament writers favor the same view, that

---

* It may be desirable to give here, in a slightly paraphrased version,
but strictly in accordance with the views of the best expositors, the essential part of the passage in Hebrews, chap. iv.:

"For God hath spoken in a certain place" (Gen. ii., 2) of the seventh
day in this wise—'And God did rest on the seventh day from all his
works;' and in this place again—'They shall not enter into my rest'
(Psa. xcv., 11).   Seeing, therefore, it still remaineth that some enter therein,
and they to whom it (God's Sabbatism) was first proclaimed entered not in,
because of disobedience (in the fall, and afterward in the sin of the Israelites in the desert), again he fixes a certain day, saying in David's writings, long after the time of Joshua—'To-day, if ye hear his voice, harden
not your hearts.'   For if Joshua had given them rest in Canaan, he would
not afterward have spoken of another day.   There is therefore yet reserved
a keeping of a Sabbath for the people of God.   For he that is entered into
his rest (that is, Jesus Christ, who has finished his work and entered into
his rest in heaven), he himself also rested from his own works, as God did
from his own.   Let us therefore earnestly strive to enter into that rest."

It is evident that in this passage God's Sabbatism, the rest intended for
man in Eden and for Israel in Canaan, Christ's rest in heaven after finish-
ing his work, and the final heavenly rest of Christ's people, are all indefinite periods mutually related, and can not possibly be natural days.

this view is essential to the true significance of the Sabbath and the Lord's day, and that it is sustained by Christ and his apostles, there is surely no need for our clinging to a mediæval notion which has no theological value, and is in opposition to the facts of nature. On the contrary, should not even children be taught these grand truths, and led to contemplate the great work of Him who is from æon to æon, and to think of that Sabbatism which he prepared for us, and which he still offers to us in the future, in connection with the succession of worlds in time revealed by geology, and which rivals in grandeur and perhaps exceeds in interest the extension of worlds in space revealed by astronomy. In truth, we should bear in mind that the great revelations of astronomy have too much habituated us to think of space-worlds rather than time-worlds, while the latter idea was evidently dominant with the Biblical writers as it is also with modern geologists. Viewed as æons—divine days, or time-worlds—the days of creation are thus a reality for all ages ; and connect themselves with the highest moral teachings of the Bible in relation to the fall of man and God's plan for his restoration, begun in this seventh æon of the world's long history, and to be completed in that second divine Sabbatism, secured by the work of redemption, the final "rest" of the "new heavens and new earth," which remains for the people of God.

But supposing that the inspired writer intended to say that the world was formed in six long periods of time, could not he have used some other word than *yom* that would have been liable to fewer doubts. There are words which might have been used, as, for instance, *eth*, time, season, or *olam*, age, ancient time, eternity. The former, however, has about it a want of precision as to its beginning and end which unfits it for this use ; the latter we have already seen is used as equivalent to the creative *yom.*   On the whole, I am unable to find any instance which would justify me in affirming that, on the supposition that Moses intended long periods, he could have better expressed the idea than by the use of the word *yom,* more especially if he and those to whom he wrote were familiar with the thought, preserved to us in the mythology of the Hindoos and Persians, and probably widely diffused in ancient Asia, that a working day of the Creator immeasurably transcends a working day of man.*

Many objections to the view which I have thus endeavored to support from internal evidence will at once occur to every intelligent reader familiar with the literature of this subject. I shall now attempt to give the principal of these objections a candid consideration.

(1.) It is objected that the time occupied in the work of creation is given as a reason for the observance of the seventh day as a Sabbath ; and that this requires us to view the days of creation as literal days.   " For in six days Jehovah made

---

* For the benefit of those who may value ancient authorities in such matters, and to show that such views may rationally be entertained independently of geology, I quote the following passage from Origen : " Cuinam quæso sensum habenti convenienter videbitur dictum, quod dies prima et secunda et tertia, in quibus et vespera nominatur, et mane, fuerint sine sole, et sine luna et sine stellis : prima autem dies sine cœlo." So St. Augustine expressly states his belief that the creative days could not be of the ordinary kind : " Qui dies, cujusmodi sint, aut perdifficile nobis, aut etiam impossibile est cogitare, quanto magis discere." Bede also remarks, "Fortassis hic diei nomen, totius temporis nomen est, et omnia volumina seculorum hoc vocabulo includit." Many similar opinions of old commentators might be quoted.   It is also not unworthy of note that the cardinal number is used here, "one day" for first day ; and though the Hebrew grammarians have sought to found on this, and a few similar passages, a rule that the cardinal may be substituted for the ordinal, many learned Hebraists insist that this use of the cardinal number implies singularity and peculiarity as well as mere priority.

the heaven and the earth, the sea and all that in them is, and
rested on the seventh day ; therefore Jehovah blessed the
Sabbath-day and sanctified it." The argument used here is,
however, as we have already seen, one of analogy. Because
God rested on his seventh day, he blessed and sanctified it,
and required men in like manner to sanctify their seventh
day.* Now, if it should appear that the working day of God
is not the same with the working day of man, and that the
Sabbath of God is of proportionate length to his working day,
the analogy is not weakened ; more especially as we find the
same analogy extended to the seventh year. If it should be
said, God worked in the creation of the world in six long ages,
and rested on the seventh, therefore man, in commemoration
of this fact, and of his own loss of an interest in God's rest
by the fall, shall sanctify the seventh of his working days, the
argument is stronger, the example more intelligible, than on
the common supposition. This objection is, in fact, a piece
of pedantic hyperorthodoxy which has too long been handed
about without investigation. I may add to what has been
already said in reference to it, the following vigorous thrust
by Hugh Miller :†

"I can not avoid thinking that many of our theologians attach a too narrow meaning to the remarkable reason attached
to the fourth commandment by the divine Lawgiver. "God
rested on the seventh day," says the text, "from all his work
which he had created and made ; and God blessed the seventh
day and sanctified it." And such is the reason given in the
Decalogue why man should rest on the Sabbath-day. God

---

* It is to be observed, however, that on the so-called literal day hypothesis the first Sabbath was not man's seventh day, but rather his first, since
he must have been created toward the close of the sixth day.

† " Footprints of the Creator."

rested on the Sabbath-day and sanctified it; and therefore man ought also to rest on the Sabbath and keep it holy. But I know not where we shall find grounds for the belief that the Sabbath-day during which God rested was merely commensurate with one of the Sabbaths of short-lived man—a brief period measured by a single revolution of the earth on its axis. We have not, as has been shown, a shadow of evidence that he resumed his work of creation on the morrow; the geologist finds no trace of post-Adamic creation; the theologian can tell us of none. God's Sabbath of rest may still exist; the work of redemption may be the work of his Sabbath-day. That elevatory process through successive acts of creation, which engaged him during myriads of ages, was of an ordinary week-day character; but when the term of his moral government began, the elevatory process peculiar to it assumed the divine character of the Sabbath. This special view appears to lend peculiar emphasis to the reason embodied in the commandment. The collation of the passage with the geologic record seems, as if by a species of retranslation, to make it enunciate as its injunction, "Keep this day, not merely as a day of memorial related to a past fact, but also as a day of co-operation with God in the work of elevation, in relation both to a present fact and a future purpose." "God keeps his Sabbath," it says, "in order that he may save; keep yours also that ye may be saved." It serves besides to throw light on the prominence of the Sabbatical command, in a digest of law of which no jot or tittle can pass away until the fulfillment of all things. During the present dynasty of probation and trial, that special work of both God and man on which the character of the future dynasty depends is the Sabbath-day work of saving and being saved.

"The common objection to that special view which regards

the days of creation as immensely protracted periods of time, furnishes a specimen, if not of reasoning in a circle, at least of reasoning from a mere assumption. It first takes for granted that the Sabbath-day during which God rested was a day of but twenty-four hours, and then argues from the supposition that, in order to keep up the proportion between the six previous working days and the seventh day of rest, which the reason annexed to the fourth commandment demands, these previous days must also have been twenty-four hours each. It would, I have begun to suspect, square better with the ascertained facts, and be at least equally in accordance with Scripture, to reverse the process, and argue that because God's working days were immensely protracted periods, his Sabbath also must be an immensely protracted period. The reason attached to the law of the Sabbath seems to be simply a reason of proportion : the objection to which I refer is an objection palpably founded on considerations of proportion, and certainly were the reason to be divested of proportion, it would be divested also of its distinctive character as a reason. Were it as follows, it could not be at all understood : "Six days shalt thou labor, etc.; but on the seventh day shalt thou do no labor, etc. ; for in six immensely protracted periods of several thousand years each did the Lord make the heavens and the earth, etc.; and then rested during a brief day of twenty-four hours ; therefore the Lord blessed the brief day of twenty-four hours and hallowed it." This, I repeat, would not be reason. All, however, that seems necessary to the integrity of the reason, in its character as such, is that the proportion of six parts to seven should be maintained. God's periods may be periods expressed algebraically by letters symbolical of unknown quantities, and man's periods by letters symbolical of quantities well known; but

if God's Sabbath be equal to one of his six working days,
and man's Sabbath equal to one of his six working days, the
integrity of proportion is maintained."

Not only does this view of the case entirely remove the objection, but, as we have already seen, it throws a new light
on the nature and reason of the Sabbath.   No good reason,
except that of setting an example, can be assigned for God's
resting for a literal day.   But if God's Sabbath of rest from
natural creation is still in progress, and if our short Sabbaths
are symbolical of the work of that great Sabbath in its present gray morning and in its coming glorious noon, then may
the Christian thank this question, incidentally raised by geology and its long periods, for a ray of light which shines
along the whole course of Scripture history, from the first
Sabbath up to that final "rest which remaineth for the people of God."*

(2.) It is objected that evening and morning are ascribed to
the first day.   This has been already noticed; it may here
be considered more fully.   The word evening in the original
is literally the darkening, the sunset, the dusk.   Morning is
the *opening* or *breaking forth* of light—the daybreak.   It must
not be denied that the explanation of these terms is attended
with some difficulty, but this is not at all lessened by narrow-
ing the day to twenty-four hours.   The first operation of the
first day was the creation of light; next we have the Creator
contemplating his work and pronouncing it to be good; then
we have the separation of the light and darkness, previously,
it is to be presumed, intermixed; and all this without the
presence of a sun or other luminary.   Which of these opera-

* This idea occurs in Lord Bacon's "Confession of Faith," and De Luc
also maintains that the Creator's Sabbath must have been of long continuance.

tions occupied the evening, and which the morning, if the day consisted of but twenty-four hours, beginning, according to Hebrew custom, in the evening? Was the old primeval darkness the evening or night, and the first breaking forth of light morning? This is almost the only view compatible with the Hebrew civil day beginning at evening, but it would at once lengthen the day beyond twenty-four hours, and contradict the terms of the record. Again, were the separated light and darkness the morning and evening? If so, why is the evening mentioned first, contrary to the supposed facts of the case? why, indeed, are the evening and morning mentioned at all, since on that supposition this is merely a repetition? Lastly, shall we adopt the ingenious expedient of dividing the evening and morning between two days, and maintaining that the evening belongs to the first and the morning to the second day, which would deprive the first day of a morning, and render the creative days, whatever their length, altogether different from Hebrew natural or civil days? It is unnecessary to pursue such inquiries farther, since it is evident that the terms of the record will not agree with the supposition of natural evening and morning. This is of itself a strong presumption against the hypothesis of civil days, since the writer was under no necessity so to word these verses that they would not give any rational or connected sense on the supposition of natural evening and morning, unless he wished to be otherwise understood.

But what is the meaning of evening and morning, if these days were long periods? Here fewer difficulties meet us. First: It is readily conceivable that the beginning and end of a period named a day should be called evening and morning. But what made the use of these divisions necessary or appropriate? I answer that nature and revelation both give grounds at least to suspect that the evening, or earlier part of each period, was a time of comparative inaction, sometimes even of retrogression, and that the latter part of each period was that of its greatest activity and perfection.   Thus, on the views stated in a former chapter, in the first day there was a time when luminous matter, either gradually concentrating itself toward the sun, or surrounding the earth itself, shed a dim but slowly increasing light; then there were day and night, the light increasing in intensity as, toward the end of the period, the luminous matter became more and more concentrated around the sun.   So in our own seventh day, the earlier part was a time of deplorable retrogression, and though the Sun of Righteousness has arisen, we have seen as yet only a dim and cloudy morning.   On the theory of days of vision, as expounded by Hugh Miller, in the "Testimony of the Rocks," in one of his noblest passages, the evening and night fall on each picture presented to the seer like the curtain of a stage.   Secondly: Though the explanation stated above is the most probable, the hypothesis of long periods admits of another, namely, that the writer means to inform us that evening and morning, once established by the separation of light from darkness, continued without cessation throughout the remainder of the period—rolling from this time uninterruptedly around our planet, like the seal cylinder over the clay.*   This explanation is, however, less applicable to the following days than to the first.   Nor does this accord with the curious fact that the seventh day, which, on the hypothesis of long periods, is still in progress, is not said to have had an evening or morning.

(3.) It is objected that the first chapter of Genesis " is not a

---

* See the quotation from Job, *supra.*

poem nor a piece of oratorical diction," but a simple prosaic narrative, and consequently that its terms must be taken in a literal sense. In answer to this, I urge that the most truly literal sense of the word, namely, the *natural* day, is excluded by the terms of the narrative; and that the word may be received as a literal day of the Creator, in the sense of one of his working periods, without involving the use of poetical diction, and in harmony with the wording of plain prosaic passages in other parts of the Bible. Examples of this have already been given. It is, however, true that, though the first chapter of Genesis is not strictly poetical, it is thrown into a metrical form which admits of some approach to a figurative expression in the case of a term of this kind.

(4.) It has been urged that in cases where day is used to denote period, as in the expressions "day of calamity," etc., the adjuncts plainly show that it can not mean an ordinary day. In answer to this, I merely refer to the internal evidence already adduced, and to the deliberate character of the statements, in the manner rather of the description of processes than of acts. The difficulties attending the explanation of the evening and the morning, and the successive creation of herbivorous and carnivorous animals, are also strong indications which should serve here to mark the sense, just as the context does in the cases above referred to.

(5.) In Professor Hitchcock's valuable and popular "Religion of Geology," I find some additional objections, which deserve notice as specimens of the learned trifles which pass current among writers on this subject, much to the detriment of sound Scriptural literature. I give them in the words of the author. 1. "From Genesis ii., 5 compared with Genesis i., 11 and 12, it seems that it had not rained on the earth till the third day; a fact altogether probable if the days were of

twenty-four hours, but absurd if they were long periods." It
strikes us that the absurdity here is all on the side of the
short days. Why should any prominence be given to a fact
so common as the lapse of two ordinary days without rain,
more especially if a region of the earth and not the whole is
referred to, and in a document prepared for a people residing
in climates such as those of Egypt and Palestine. But what
could be more instructive and confirmatory of the truth of the
narrative than the fact that in the two long periods which preceded the formation and clearing up of the atmosphere or
firmament, on which rain depends, and the elevation of the
dry land, which so greatly modifies its distribution, there had
been no rain such as now occurs. This is a most important
fact, and one of the marked coincidences of the record with
scientific truth. The objection, therefore, merely shows that
the ordinary day hypothesis tends to convert one of the finest
internal harmonies of this wonderful history into an empty
and, in some respects, absurd commonplace. 2. "This hypothesis (that days are long periods) assumes that Moses describes the creation of all the animals and plants that have
ever lived on our globe. But geology decides that the species now living, since they are not found in the rocks any
lower than man is,* could not have been contemporaneous
with those in the rocks, but must have been created when
man was—that is, in the sixth day. Of such a creation no
mention is made in Genesis; the inference is that Moses does
not describe the creation of the existing races, but only of
those that lived thousands of years earlier, and whose exist-

---

* This is not strictly correct, as many animals, especially of the lower
tribes, extend back to the early tertiary periods, long before the creation
of man ; a fact which of itself is irreconcilable with the Mosaic narrative
on the theory of literal or ordinary days.

ence was scarcely suspected till modern times. Who will admit such an absurdity?" In answer to this objection, I remark that it is based on a false assumption. The hypothesis of long periods does not require us to assume that Moses notices all the animals and plants that have ever lived, but on the contrary that he informs us only of the *first appearance* of each great natural type in the animal and vegetable kingdoms ; just as he informs us of the first appearance of dry land on the third day, but says nothing of the changes which it underwent on subsequent days. Thus plants were created on the third day, and though they may have been several times destroyed and renewed as to genera and species, we infer that they continued to exist in all the succeeding days, though the inspired historian does not inform us of the fact. So also many tribes of animals were created in the early part of the fifth day, and it is quite unnecessary for us to be informed that these tribes continued to exist through the sixth day. If the days were long periods, the inspired writer could not have adopted any other course, unless he had been instructed to write a treatise on Palæontology, and to describe the fauna and flora of each successive period with their characteristic differences. 3. "Though there is a general resemblance between the order of creation as described in Genesis and by geology, yet when we look at the details of the creation of the organic world, as required by this hypothesis, we find manifest discrepancy. Thus the Bible represents plants only to have been created on the third day, and animals not till the fifth ; and hence at least the lower half of the fossiliferous rocks ought to contain nothing but vegetables. Whereas in fact the lower half of these rocks, all below the carboniferous, although abounding in animals, contain scarcely any plants, and these in the lowest strata fucoids or sea-weeds.

But the Mosaic account evidently describes flowering and seed-bearing plants, not flowerless and seedless algæ. Again, reptiles are described in Genesis as created on the fifth day; but reptilia and batrachians existed as early as the time when the lower carboniferous and even old red sandstone were in course of deposition, as their tracks on those rocks in Nova Scotia and Pennsylvania evince.* In short, if we maintain that Moses describes fossils as well as living species, we find discrepancy instead of correspondence between his order of creation and that of geology." In this objection it is assumed that the geological history of the earth goes back to the third day of creation, or, in other words, to the dawn of organic life. None of the greater authorities in geology would, however, now venture to make such an assertion, and the progress of geology is rapidly making the contrary more and more probable. The fact is that, on the supposition that the days of creation are long periods, the whole series of the fossiliferous rocks belongs to the fifth and sixth days; and that for the early plant creation of the third day, and the great physical changes of the fourth, geology has nothing as yet to show, except a mass of metamorphosed eozoic rocks which have hitherto yielded no fossils except a few Protozoa; but which contain vast quantities of carbon in the form of graphite, which may be the remains of plants.

I have much pleasure in quoting, as a further answer to these objections, the following from Professor Dana :†

"Accepting the account in Genesis as true, the seeming

---

* Since this was written, the bones of many Batrachian reptiles have been found in the Carboniferous, both in Europe and America. No reptilian remains have yet been found in the Devonian rocks.

† *Biblical Repository*, 1856. See also an excellent paper by Prof. C. H. Hitchcock, *Bibliotheca Sacra*, 1867.

discrepancy between it and geology rests mainly here : Geology holds, and has held from the first, that the progress of creation was mainly through secondary causes; for the existence of the science presupposes this. Moses, on the contrary, was thought to sustain the idea of a simple fiat for each step. Grant this first point to science, and what farther conflict is there? *The question of the length of time*, it is replied. But not so ; for if we may take the record as allowing more than six days of twenty-four hours, the Bible then places no limit to time. *The question of the days and periods*, it is replied again. But this is of little moment in comparison with the first principle granted. Those who admit the length of time and stand upon days of twenty-four hours have to place geological time *before* the six days, and then assume a chaos and reordering of creation, on the six-day and fiat principle, after a previous creation that had operated for a long period through secondary causes. Others take days as periods, and thus allow the required time, admitting that creation was one in progress, a grand whole, instead of a *first* creation excepting man by one method, and a *second* with man by the other. This is now the remaining question between the theologians and geologists ; for all the minor points, as to the exact interpretation of each day, do not affect the general concordance or discordance of the Bible and science.

"On this point geology is now explicit in its decision, and indeed has long been so. It proves that there was no return to chaos, no great revolution, that creation was beyond doubt one in its progress. We know that some geologists have taken the other view. But it is only in the capacity of theologians, and not as geologists. The Rev. Dr. Buckland, in placing the great events of geology between the first and second verses of the Mosaic account, did not pretend that there was a geological basis for such an hypothesis; and no writer since has ever brought forward the first fact in geology to support the idea of a rearrangement just before man; not one solitary fact has ever been appealed to. The conclusion was on Biblical grounds, and not in any sense on geological. The best that Buckland could say, when he wrote twenty-five years since, was that geology did not absolutely disprove such an hypothesis; and that can not be said now.

"It is often asserted, in order to unsettle confidence in these particular teachings of geology, that geology is a changing science. In this connection the remark conveys an erroneous impression. Geology is a progressive science; and all its progress tends to establish more firmly these two principles: (1) The slow progress of creation through secondary causes, as explained; and (2) the progress by periods analogous to the days of Genesis."

I have, I trust, shown that the principal objections to the lengthening of the Mosaic days into great cosmical periods are of a character too light and superficial to deserve any regard. I shall now endeavor to add to the internal evidence previously given some considerations of an external character which support this view.

1. The fact that the creation was progressive, that it proceeded from the formation of the raw material of the universe, through successive stages, to the perfection of living organisms, if we regard the analogy of God's operations as disclosed in the geological history of the earth and in the present course of nature, must impress us with a suspicion that long periods were employed in the work. God might have prepared the earth for man in an instant. He did not choose to do so, but on the contrary proceeded step by step; and the record he has given us does not receive its full significance

nor attain its full harmony with the course of geological history, unless we can understand each day of the creative week as including a long succession of ages.

2. We have, as already explained, reason to believe that the seventh day at least has been of long duration. At the close of the sixth, God rested from all his work of material creation, and we have as yet no evidence that he has resumed it. Neither theologians nor evolutionists will, I presume, desire to maintain that any strictly creative acts have occurred in the modern period of geology. We know that the present day, if it is the seventh, has lasted already for at least six thousand years, and, if we may judge from the testimony of prophecy, has yet a long space to run, before it merges in that "new heaven and new earth" for which all believers look, and which will constitute the first day of an endless sabbatism.

3. The philosophical and religious systems of many ancient nations afford intimations of the somewhat extensive prevalence in ancient times of the notion of long creative periods, corresponding to the Mosaic days. These notions, in so far as they are based on truth, are probably derived from the Mosaic narrative itself, or from the primitive patriarchal documents which may have formed the basis of that narrative. They are, no doubt, all more or less garbled versions, and can not be regarded as of any authority, but they serve to show what was the interpretation of the document in a very remote antiquity. I have collected from a variety of sources the following examples:

The ancient mythology of Persia appears to have had six creative periods, each apparently of a thousand years, and corresponding very nearly with the Mosaic days.* The

* Rhode, quoted by McDonald, "Creation and the Fall," p. 62; Eusebius, Chron. Arm.

Chaldeans had a similar system, to which in a previous chapter we have already referred. The Etruscans possessed a history of the creation, somewhat resembling that of the Bible, and representing the creation as occupying six periods of a thousand years each.*

The Egyptians believed that the world had been subject to a series of destructions and renewals, the intervals between which amounted to 120,000 years, or, according to other authorities, to 300,000 or 360,000 years. This system of destruction and renewal the Egyptian priests appear to have wrought out into considerable detail, but though important truths may be concealed under their mysterious dogmas, it will not repay us to dwell on the fragments that remain of them. There can be no doubt, however, that at least the basis of the Egyptian cosmogony must have been the common property of all the Hamite nations, of which Egypt was the greatest and most permanent; and therefore in all probability derived from the ideas of creation which were current not long after the Deluge. The Egyptians appear also, as already stated, to have had a physical cosmogony, beginning with a chaos in which heaven and earth were mingled, and from which were evolved fiery matters which ascended into the heavens, and moist earthy matters which formed the earth and the sea; and from these were produced, by the agency of solar heat, the various animals. The terms of this cosmogony, as it is given by Diodorus Siculus, indicate the belief of long formative periods.†

The Hindoos have a somewhat extended, though, according to the translations, a not very intelligible cosmogony. It

---

* Suidas, Lexicon—" Tyrrenia."
† Diodorus Siculus, bk. i.   Prichard, Egyptian Mythology.

plainly, however, asserts long periods of creative work, and is interesting as an ancient cosmogony preserved entire and without transmission through secondary channels. The following is a summary, in so far as I have been able to gather it, from the translation of the Institutes of Menu by Sir W. Jones.*

The introduction to the Institutes represents Menu as questioned by the "divine sages" respecting the laws that should regulate all classes or castes. He proceeds to detail the course of creation, stating that the "Self-existing Power,† undiscovered, but making this world discernible, He whom the mind alone can perceive, whose essence eludes the external senses, who has no visible parts, who exists from eternity, even the soul of all being, whom no being can comprehend, shone forth in person."

After giving this exalted view of the Creator, the writer proceeds to state that the Self-existent created the waters, and then an egg, from which he himself comes forth as Brahma the forefather of spirits. "The waters are called Nara because they are the production of *Nara*, the spirit of God, and since they were his first *Ayana*, or place of motion, he thence is named *Narayana*, or moving on the waters. In the egg Brahma remained a year,' and caused the egg to divide, forming the heaven above and the earth beneath, and the subtile ether, the eight regions, and the receptacle of waters between. He then drew forth from the supreme soul mind with all its powers and properties." The rest of the account appears to be very confused, and I confess to a great extent unintelli-

---

* "Asiatic Researches."
† This name is exactly identical in meaning with the Hebrew Jehovah

gible to me. There follows, however, a continuation of the narrative, stating that there is a succession of seven Menus, each of whom produces and supports the earth during his reign. It is in the account of these successive Menus that the following statement respecting the days and years of Brahma occurs :

"A day of the Gods is equal to a year. Four thousand years of the Gods are called a Critya or Satya age. Four ages are an age of the Gods. *One thousand divine ages (equal to more than four millions of human years) are a day of Brahma the Creator.* Seventy-two divine ages are one manwantara. * * * The aggregate of four ages they call a divine age, and believe that in every thousand such ages, or in every day of Brahma, fourteen Menus are successively invested with the sovereignty of the earth. Each Menu they suppose transmits his authority to his sons and grandsons during a period of seventy-two divine ages, and such a period they call a manwantara. Thirty such days (of the Creator), or calpas, constitute a month of Brahma; twelve such months one of his years, and 100 such years his age, of which they assert that fifty years have elapsed. We are thus, according to the Hindoos, in the first day or calpa of the fifty-first year of Brahma's life, and in the twenty-eighth divine age of the *seventh manwantara* of that day. In the present day of Brahma the first Menu was named the Son of the Self-existent, and by him the institutes of religion and civil duties are said to have been delivered. In his time occurred a new creation called the *Lotos* creation." Of five Menus who succeeded him, Sir William could find little but the names, but the accounts of the seventh are very full, and it appears that in his reign the earth was destroyed by a flood. Sir William suggests that the first Menu may represent the creation, and that the seventh may be Noah. The name Menu or Manu is equivalent to "man," and signifies "the intelligent."*

In this Hindoo cosmogony we have many points of correspondence with the Scripture narrative : for instance, the Self-existent Creator ; the agency of the Son of God and the Holy Spirit ; the absolute creation of matter ; the hovering of the Spirit over the primeval waters ; the sevenfold division of the creative process ; and the idea of days of the Creator of immense duration. If we suppose the day of Brahma in the Hindoo cosmogony to represent the Mosaic day, then it amounts to no less than 4,320,000 years ; or if, with Sir W. Jones, we suppose the manwantara to represent the Mosaic day, its duration will be 308,571 years ; and the total antiquity of the earth, without counting the undefined "beginning," will be either more than twenty-five or than two millions of years. It would be folly, however, to suppose that these Hindoo numbers, which are probably purely conjectural, or based on astronomical cycles, make any near approximation to the facts of the case. The Institutes of Menu are probably in their present form not of great antiquity, but there are other Hindoo documents of greater age which maintain similar views, and it is probable that the account of the creation in the Institutes is at least an imperfect version of the original narrative as it existed among the earliest colonists of India.† It corresponds in many points with the oldest notions on these subjects that remain to us in the wrecks of the mythology of Egypt and other an-

---

* Muller, Sanscrit Literature.

† The theology of the Institutes is clearly primitive Semitic in its character ; and therefore, if the Bible is true, must be older than the Aryan theogony of the Rig-Veda, as expounded by Muller, whatever the relative age of the documents.

cient nations, and it aids in proving that the fabulous ages of
gods and demi-gods in the ancient mythologies *are really pre-
Adamite;* and belong not to human history, but to the work of
creation. It also shows that the idea of long creative periods
as equivalents of the Mosaic days must, in the infancy of the
postdiluvian world, have been very widely diffused. Such
evidence is, no doubt, of small authority in the interpretation
of Scripture ; but it must be admitted that serious consideration is due to a method of interpretation which thus tends to
bring the Mosaic account into harmony with the facts of
modern science, and with the belief of almost universal antiquity, and at the same time gives it its fullest significance
and most perfect internal symmetry of parts. It is also very
interesting to note the wide diffusion among the most ancient
nations of cosmological views identical in their main features
with those of the Bible, proving, almost beyond doubt, that
these views had some common and very ancient source, and
commanded universal belief among the primitive tribes of men.

I have hitherto in this part of the discussion avoided detailed reference to what may be regarded as the " prophetic
day " view of the narrative of creation. This may be shortly
stated as follows : In the prophetical parts of Scripture the
prophet sees in vision, as in a picture or acted scene, the
events that are to come to pass, and in consequence represents years or longer periods by days of vision. Now the
revelation of the pre-Adamite past is in its nature akin to
that of the unknown future ; and Moses may have seen these
wondrous events in vision—in visions of successive days—
under the guise of which he presents geological time. Some
things in the form of the narrative favor this view, and it certainly affords the most clearly intelligible theory as to the
mode in which such a revelation may have been made to

man. It is advocated by Kurtz, by the author of an excellent little work, the "Harmony of the Mosaic and Geological Records," by Hugh Miller, and more recently by Tayler Lewis. To these writers I must refer for its more full illustration, and for the grand pictorial view which it gives of the vision of the creative week.

In reviewing the somewhat lengthy train of reasoning into which the term "day" has led us, it appears that from internal evidence alone it can be rendered probable that the day of creation is neither the natural nor the civil day. It also appears that the objections urged against the doctrine of day-periods are of no weight when properly scrutinized, and that it harmonizes with the progressive nature of the work, the evidence of geology, and the cosmological notions of ancient nations. I do not suppose that this position has been incontrovertibly established; but I believe that every serious difficulty has been removed from its acceptance; and with this, for the present, I remain satisfied. Every step of our subsequent progress will afford new criteria of its truth or fallacy.

One further question of some interest is—What, according to the theory of long creative days and the testimony of geology, would be the length and precise cosmical nature of these days? With regard to the first part of the question, we do not know the actual value of our geological ages in time; but it is probable that each great creative æon may have extended through millions of years. As to the nature of the days, this may have been determined by direct volitions of the Creator, or indirectly by some of those great astronomical cycles which arise from the varying eccentricity of the earth's orbit, or the diminution of the velocity of its rotation, or by its gradual cooling.

With reference to these points, science has as yet little in-

formation to give. Sir William Thomson has, indeed, indicated for the time since the earth's crust first began to form
a period of between one and two hundred millions of years;
but Professor Guthrie Tait, on the other hand, argues that
ten or fifteen millions of years are probably sufficient,* and
Lockyer has suggested an hypothesis of successive rekind-
lings of the solar heat which might give a more protracted
time than that of Thomson. Some of the hypotheses of derivation current, but which are based rather on philosophical
speculation than on scientific fact, would also require a
longer time than that allowed by Thomson ; and it is to be
regretted· that some geologists, by giving credence to such
hypotheses of derivation, and by loose reasoning on the time
required for the denudation and deposition of rocks, have
been induced to commit themselves to very extravagant estimates as to geological time. On the whole, it is evident that
only the most vague guesses can at present be based on the
facts in our possession, though the whole time required has
unquestionably been very great, the deposition of the series
of stratified rocks probably requiring at least the greater part
of the minimum time allowed by Thomson.†

As to the cosmical nature of the periods, while some geologists appear to regard the whole of geological time as a continuous evolution without any breaks, it is evidently more in
accordance with facts to hold that there have been cycles of
repose and activity succeeding each other, and that these
have been of different grades. In the succession of deposits
it is plain that periods of depression and upheaval common
to all the continental masses have succeeded each other at

* " Recent Advances in Physical Science."
† Croll's " Climate and Time " contains some interesting facts as to this.

somewhat regular intervals, and that within these periods there have been alternations of colder and warmer climates. These, however, are not equal to the creative days of our record, for they are greatly more numerous. They are but the vastly protracted hours of these almost endless days. Beyond and above these there is another grade of geological period, marked not by mere gradual elevation and depression of the continental areas, but by vast crumplings of the earth's crust and enormous changes of level. Such a great movement unquestionably closed the Eozoic period of geology. Another of less magnitude occurred in what is termed the Permian age at the end of the Palæozoic. A third terminated the Mesozoic age, and introduced the Tertiary or Kainozoic. Perhaps we should reckon the glacial age, though characterized by far less physical change than the others, as a fourth. The possible physical causes which have been suggested for such greater disturbances are the collapses of the crust in equatorial regions, which may be supposed to have resulted at long intervals of time, from the gradual retardation of the earth's rotation caused by the tides, or the similar collapses and other changes due to the shrinkages of the earth's interior caused by its gradual cooling, and to the unequal deposition of material by water on different parts of its surface.* The more full discussion of these points belongs, however, to a future chapter.

These greater movements of the crust, would, as already stated, coincide to some extent with the later creative days in the manner indicated below :

Collapse of crust at close of Eozoic Time,	Close of Fourth Æon, and beginning of Fifth.

---

* See the discussion of this in the author's "Story of the Earth," and in Sir William Thomson's British Association Address, 1876.

Collapse in Permian Period and end of Palæozoic Time,	} Middle of Fifth Æon.
Great subsidence and collapse at close of Mesozoic Age,	} Close of Fifth Æon, and beginning of Sixth.
Great subsidence of the Pleistocene or Glacial Age,	} End of Sixth Æon.

The question recurs—Why are God's days so long? He is not like us, a being of yesterday. He is "from Olam to Olam," and even in human history one day is with him as a thousand years; and we who live in these later days of the world know full well how slow the march of his plan has been even in human history. We shall know in the endless ages of a future eternity that even to us these long creative days may at last become but as watches in the night.

CHAPTER VII.

THE ATMOSPHERE.

"And God said, Let there be an expanse between the waters; and let it separate the waters from the waters. And God made the expanse, and separated the waters which are under the expanse from the waters which are over the expanse : and it was so. And God called the expanse Heaven. And the evening and the morning were the second day."—Genesis i. 6–8.

AT the opening of the period to which we are now introduced the earth was covered by the waters, and these were in such a condition that there was no distinction between the seas and the clouds. No atmosphere separated them, or, in other words, dense fogs and mists everywhere rested on the surface of the primeval ocean. To understand as far as possible the precise condition of the earth's surface at this period, it will be necessary to notice the present constitution of the atmosphere, especially in its relations to aqueous vapor.

The regular and constant constituents of the atmosphere are the elements oxygen and nitrogen, which, at the temperature and pressure existing on the surface of our globe, are permanently aeriform or gaseous. Beside these gases, the air always contains a quantity of the vapor of water in a perfectly aeriform and transparent condition. This vapor is not, however, permanently gaseous. At all temperatures below 212 degrees it tends to the liquid state ; and its elastic force, which preserves its particles in the separated state of

vapor, increases or diminishes at a more rapid rate than the
increase or diminution of temperature. Hence the quantity
of vapor that can be suspended in clear air depends on the
temperature of the air itself. As the temperature of the air
rises, its power of sustaining vapor increases more rapidly
than its temperature ; and as the temperature of the air falls,
the elastic force of its contained vapor diminishes in a greater ratio, until it can exist as an invisible vapor no longer, but
becomes condensed into minute bubbles or globules, forming
cloud, mist, or rain. Two other circumstances operate along
with these properties of air and vapor. The heat radiated
from the earth's surface causes the lower strata of air to be,
in ordinary circumstances, warmer than the higher ; and, on
the other hand, warm air, being lighter than that which is
colder, the warm layer of air at the surface continually tends
to rise through and above the colder currents immediately
over it. Let us consider the operation of the causes thus
roughly sketched in a column of calm air. The lower portion becomes warmed, and if in contact with water takes up a
quantity of its vapor proportioned to the temperature, or in
ordinary circumstances somewhat less than this proportion.
It then tends to ascend, and as it rises and becomes mixed
with colder air it gradually loses its power of sustaining
moisture, and at a height proportioned to the diminution of
temperature and the quantity of vapor originally contained in
the air, it begins to part with water, which becomes condensed in the form of mist or cloud ; and the surface at
which this precipitation takes place is often still more distinctly marked when two masses or layers of air at different
temperatures become intermixed ; in which case, on the
principle already stated, the mean temperature produced is
unable to sustain the vapor proper to the two extremes, and

moisture is precipitated. It thus happens that layers of cloud accumulate in the atmosphere, while between them and the surface there is a stratum of clear air. Fogs and mists are in the present state of nature exceptional appearances, depending generally on local causes, and showing what the world might be but for that balancing of temperature and the elastic force of vapor which constitutes the atmospheric firmament.*

The quantity of water thus suspended over the earth is enormous. "When we see a cloud resolve itself into rain, and pour out thousands of gallons of water, we can not comprehend how it can float in the atmosphere."† The explanation is—1st, the extreme levity of the minute globules, which causes them to fall very slowly ; 2d, they are supported by currents of air, especially by the ascending currents developed both in still air and in storms ; 3dly, clouds are often dissolving on one side and forming on another. A cloud gradually descending may be dissolving away by evaporation at the base as fast as new matter is being added above. On the other hand, an ascending warm current of air may be constantly depositing moisture at the base of the cloud, and this may be evaporating under the solar rays above. In this case a cloud is "merely the visible form of an aerial space, in which certain processes are at the moment in equilibrium, and all the particles in a state of upward movement."‡ But so soon as condensation markedly exceeds evaporation, rain falls, and the atmosphere discharges its vast load of water—how vast we may gather from the fact

---

* Daniell's Meteorological Essays ; Prout's Bridgewater Treatise ; art. "Meteorology," Encyc. Brit. ; "Maury's Physical Geography of the Sea."
† Kaemtz, "Course of Meteorology."
‡ Encyc. Brit., art. "Meteorology."

that the waters of all the rivers are but a part of the overflow-
ings of the great atmospheric reservoir. "God binds up the
waters in his thick cloud, and the cloud is not rent under
them." It is thus that the terrestrial waters are divided into
those above and those below that expanse of clear air in
which we live and move, exempt from the dense, dark mists
of the earth's earlier state, yet enjoying the benefits of the
cloudy curtain that veils the burning sun, and of the cloudy
reservoirs that drop down rain to nourish every green thing.

We have no reason to suppose that the laws which regulate
mixtures of gases and vapors did not prevail in the period in
question. It is probable that these laws are as old as the
creation of matter; but the condition of our earth up to the
second day must have been such as prevented them from
operating as at present. Such a condition might possibly be
the result of an excessive evaporation occasioned by internal
heat. The interior of the earth still remains in a heated
state, and includes large subterranean reservoirs of melted
rock, as is proved by the increase of temperature in deep
mines and borings, and by the widely extended phenomena
of hot springs and volcanic action. At the period in question
the internal temperature of the earth was probably vastly
greater than at present, and perhaps the whole interior of the
globe may have been in a state of igneous fluidity. At the
same time the external solid crust may have been thin, and it
was not fractured and thickened in places by the upheaval of
mountain chains or the deposition of great and unequal sheets
of sediment; for, as I may again remind the reader, the primitive chaos did not consist of a confused accumulation of
rocky masses, but the earth's crust must then have been more
smooth and unbroken than at any subsequent period. This
being the internal condition of the earth, it is quite conceiva-

ble, without any violation of the existing laws of nature, that the waters of the ocean, warmed by internal heat, may have sent up a sufficient quantity of vapor to keep the lower strata of air in a constant state of saturation, and to occasion an equally constant precipitation of moisture from the colder strata above. This would merely be the universal operation of a cause similar to that which now produces fogs at the northern limit of the Atlantic Gulf Stream, and in other localities where currents of warm water flow under or near to cooler air. Such a state of things is more conceivable in a globe covered with water, and consequently destitute of the dry and powerfully radiating surfaces which land presents, and receiving from without the rays, not of a solar orb, but of a comparatively feeble and diffused luminous ether. The continued action of these causes would gradually cool the earth's crust and its incumbent waters, until the heat from without preponderated over that from within, when the result stated in the text would be effected.

The statements of our primitive authority for this condition of the earth might also be accounted for on the supposition that the permanently gaseous part of the atmosphere did not at the period in question exist in its present state, but that it was on the second day actually elaborated and caused to take its place in separating the atmospheric from the oceanic waters. The first is by far the more probable view; but we may still apply to such speculations the words of Elihu, the friend of Job:

"Stand still and consider the wonderful works of God.
Dost thou know when God disposes them,
And the lightning of his cloud shines forth?
Dost thou know the poising of the dark clouds,
The wonderful works of the Perfect in knowledge?"

We may now consider the words in which this great improvement in the condition of the earth is recorded. The Hebrew term for the atmosphere is *Rakiah*, literally, something expanded or beaten out—an expanse. It is rendered in our version "firmament," a word conveying the notion of support and fixity, and in the Septuagint "*Stereoma*," a word having a similar meaning. The idea conveyed by the Hebrew word is not, however, that of *strength*, but of *extent;* or as Milton—the most accurate of expositors of these words—has it:

> "The firmament, expanse of liquid, pure,
> Transparent, elemental air, diffused
> In circuit to the uttermost convex
> Of this great round."

That this was really the way in which this word was understood by the Hebrews appears from several passages of the Bible. Job says of God, "Who alone *spreadeth* out the heavens."* David, in the 104th Psalm, which is a poetical paraphrase of the history of creation, speaks of the Creator as "*stretching* out the heavens as a curtain." In later writers, as Isaiah, Jeremiah, Ezekiel, similar expressions occur. The notion of a solid or arched firmament was probably altogether remote from the minds of these writers. Such beliefs may have prevailed at the time when the Septuagint translation was made, but I have no hesitation in affirming that no trace of them can be found in the Old Testament. In proof of this, I may refer to some of the passages which have been cited as affording the strongest instances of this kind of

---

* It is not meant that the word *rakiah* occurs in these passages, but to show how by other words the idea of stretching out or extension rather than solidity is implied. The verb in the first two passages is *nata*, to spread out.

"accommodation." In Exodus xxiv., 10, we are told, "And they saw the God of Israel, and under his feet as it were a paved work of sapphire, and as it were the heaven itself in its clearness." This is evidently a comparison of the pavement seen under the feet of Jehovah to a sapphire in its color, and to the heavens in its transparency. The intention of the writer is not to give information respecting the heavens, or to liken them either to a pavement or a sapphire ; all that we can infer is that he believed the heavens to be clear or transparent. Job mentions the "pillars of heaven," but the connection shows that this is merely a poetical expression for lofty mountains. The earthquake causes these pillars of heaven to "tremble." We are informed in the book of Job that God "ties up his waters in his thick cloud, and the cloud is not rent under them." We are also told of the "treasures of snow and the treasures of hail," and rain is called the "bottles of heaven," and is said to be poured out of the "lattices of heaven." I recognize in all these mere poetical figures, not intended to be literally understood. Some learned writers wish us to believe that the intention of the Bible in these places is actually to teach that the clouds are contained in skin bottles, or something similar, and that they are emptied through hatches in a solid firmament. To found such a belief, however, on a few figurative statements, seems ridiculous, especially when we consider that the writers of the Scriptures show themselves to be well acquainted with nature, and would not be likely on any account to deviate so far from the ordinary testimony of the senses; more especially as by doing so they would enable every unlettered man who has seen a cloud gather on a mountain's brow or dissolve away before increasing heat to oppose the evidence of his senses to their statements, and perhaps to reject them with scorn as a bare-

faced imposture. But, lastly, we are triumphantly directed to
the question of Elihu in his address to Job :

> "Hast thou with him stretched out the sky,
> Which is firm and like a molten mirror?"

But the word translated sky here is not "*rakiah*," or
"*shamayim*," but another signifying the *clouds*, so that we
should regard Elihu as speaking of the apparent firmness or
stability, and the beautiful reflected tints of the clouds. His
words may be paraphrased thus : "Hast thou aided Him in
spreading out those clouds, which appear so stable and self-
sustaining, and so beautifully reflect the sunlight?"* The
above passages form the only authority which I can find in
the Scriptures for the doctrine of a solid firmament, which
may therefore be characterized as a modern figment of men
more learned in books but less acquainted with nature than
the Scripture writers. As a contrast to all such doctrines I
may quote the sublime opening of the poetical account of
creation in Psalm civ., which we may also take here as elsewhere as the oldest and most authoritative commentary on
the first chapter of Genesis :

> "Bless the Lord, O my soul !
> O Lord, my God, thou art very great :
> Thou art clothed with honor and majesty,
> Who coverest thyself with light as with a garment,
> Who stretchest out the heavens like a curtain (of a tent),
> *Who layest the beams of thy chambers in the waters,*
> *Who makest the clouds thy chariots,*
> *Who walkest upon the wings of the wind.*"

The waters here are those above the firmament, the whole

---

* See also Humboldt, "Cosmos," vol. ii., pt. 1.

of this part of the Psalm being occupied with the heavens; and there is no place left for the solid firmament, of which the writer evidently knew nothing. He represents God as laying his chambers on the waters, instead of on the supposed firmament, and as careering in cloudy chariots on the wings of the wind, instead of over a solid arch. For all the above reasons, we conclude that the " expanse " of the verses under consideration was understood by the writers of the book of God to be *aerial*, not *solid;* and the " establishment of the clouds above," as it is finely called in Proverbs, is the effect of those meteorological laws to which I have already referred, and which were now for the first time brought into operation by the divine Legislator. The Hebrew theology was not of a kind to require such expedients as that of solid heavenly arches; it recurred at once to the will—the decree—of Jehovah; and was content to believe that through this efficient cause the " rivers run into the sea, yet the sea is not full," for " to the place whence the rivers came, thither they return again," through the agency of those floating clouds, " the waters above the heavens," which " pour down rain according to the vapor thereof."

God called the expanse " Heaven." In former chapters we have noticed that heaven in the popular speech of the Hebrews, as in our own, had different meanings, applying alike to the cloudy, the astral, and the spiritual heavens. The Creator here sanctions its application to the aerial expanse; and accordingly throughout the Scriptures it is used in this way; *rakiah* occurs very rarely, as if it had become nearly obsolete, or was perhaps regarded as a merely technical or descriptive term. The divine sanction for the use of the term heaven for the atmosphere is, as already explained, to indicate that this popular use is not to interfere with its application to the whole universe beyond our earth in verse 1st.

The poetical parts of the Bible, and especially the book of Job, which is probably the most ancient of the whole, abound in references to the atmosphere and its phenomena. I may quote a few of these passages, to enable us to understand the views of these subjects given in the Bible, and the meaning attached to the creation of the atmosphere, in very ancient periods. In Job, 38th chapter, we have the following:

> " In what way is the lightning distributed,
>   And how is the east wind spread abroad over the earth ?
>   Who hath opened a channel for the pouring rain,
>   Or a way for the thunder-flash ?
>   To cause it to rain on the land where no man is,
>   In the desert where no one dwells ;
>   To saturate the desolate and waste ground,
>   And to cause the bud of the tender herb to spring forth."

Here we have the unequal and unforeseen distribution of thunder-storms, beyond the knowledge and power of man, but under the absolute control of God, and designed by him for beneficent purposes. Equally fine are some of the following lines :

> " Dost thou lift up thy voice to the clouds,
>   That abundance of waters may cover thee ?
>   Dost thou send forth the lightnings, and they go,
>   And say unto thee, Here are we ?
>   Who can number the clouds by wisdom,
>   Or cause the bottles of heaven to empty themselves ?
>   When the dust groweth into mire,
>   And the clods cleave fast together ?"

In the 36th and 37th chapters of the same book we have a grand description of atmospheric changes in their relation to man and his works. The speaker is Elihu, who in this ancient book most favorably represents the knowledge of nature that existed at a time probably anterior to the age of Moses —a knowledge far superior to that which we find in the works of many modern poets and expositors, and accompanied by an intense appreciation of the grandeur and beauty of natural objects :

> " For he draweth up the drops of water,
>   Rain is condensed * from his vapor,
>   Which the clouds do drop,
>   And distill upon man abundantly.
>   Yea, can any understand the distribution of the clouds
>   Or the thundering of his tabernacle.†
>   Behold he spreadeth his lightning upon it,
>   He covereth it as with the depths of the sea.‡
>   By these he executes judgment on the people,
>   By these also he giveth food in abundance ;
>   His hands he covers with the lightning,
>   And commands it (against the enemy) in its striking ;
>   He uttereth to it his decree,§
>   Concerning the herd as well as proud man.
>   At this also my heart trembles,
>   And bounds out of its place ;
>   Hear attentively the thunder of his voice,
>   And the loud sound that goes from his mouth.
>   He directs it under the whole heavens,
>   And his lightning to the ends of the earth.
>   After it his voice roareth,

---

* Heb., " they refine."

† " His pavilion round about him was dark waters and thick clouds of the skies," Psa. xviii. This expression explains that in the text.

‡ Or " He darkens the depths of the sea."

§ Translation of these lines much disputed and very difficult. Gesenius and Conant render it, " His thunder tells of him ; to the herds even of him who is on high."

He thundereth with the voice of his majesty ;
And delays not (the tempest) when his voice is heard.
God thundereth marvellously with his voice,
He doeth wonders which we can not comprehend ;
For he saith to the snow, Be thou on the earth.
Also to the pouring rain, even the great rain of his might.
He sealeth up the hand of every man,
That all men may know his work.
Then the beasts go to their dens,
And remain in their caverns.
Out of the south cometh the whirlwind
And cold out of the north,
By the breath of God the frost is produced
And the breadth of waters becomes bound ;
With moisture he loads the thick cloud,
He spreads the cloud of his lightning,
And it is turned about by his direction,
To execute his pleasure on the face of the world ;
Whether for correction, for his land, or for mercy,
He causeth it to come.
Hearken unto this, O Job,
Stand still and consider the wonderful works of God.
Dost thou know when God disposes these things,
And the lightning of his cloud flashes forth ?
Dost thou know the poising of the clouds,
The wonderful work of the Perfect in knowledge ?
When thy garments become warm
When he quieteth the earth by the south wind ;
Hast thou with him spread out the clouds
Firm and like a molten mirror ?"*

---

* I take advantage of this long quotation to state that in the case of
this and other passages quoted from the Old Testament I have carefully
consulted the original ; but have availed myself freely of the renderings
of such of the numerous versions and commentaries as I have been able
to obtain, whenever they appeared accurate and expressive, and have not
scrupled occasionally to give a free translation where this seemed necessary to perspicuity.   In the book of Job, I have consulted principally the

It would not be easy to find, in the poetry of any nation or time, a description of so many natural phenomena, so fine in feeling or truthful in delineation. It should go far to dispel the too prevalent ideas of early Oriental ignorance, and should lead to a more full appreciation of these noble pictures of nature, unsurpassed in the literature of any people or time. I trust that the previous illustrations are sufficient to show, not only that the *stereoma*, or solid firmament of the Septuagint, is not to be found in Scripture, but that the positive doctrine of the Bible on the subject is of a very different character. For instance, in the above extract from the book of Job, Elihu speaks of the poising or suspension of the clouds as inscrutable, and tells us that God draws up water into the clouds, and pours down rain according to the vapor thereof; he also speaks of the clouds as being scattered before the brightness of the sun ; and notices, in truthful as well as exalted language, the nature and succession of the lightning's flash, the thunder, and the precipitation of rain that follows. Solomon also informs us that the "establishment of the clouds above" is due to the law or will of Jehovah. Finally, in this connection, the divine sanction given to the use of the term heaven for the atmosphere may in itself be regarded as an intimation that no definite barrier separates our film of atmosphere from the boundless abyss of heaven without.

Of this period natural science gives us no intimation. In the earliest geological epochs organic life, dry land, and an atmosphere already existed. At the period now under consideration the two former had not been called into existence, and the latter was in process of elaboration from the materi-

translation appended to Barnes's Commentary, Conant's translation, 1857, and those of Tayler Lewis and Evans in Schaff's edition of Lange, 1874.

H

als of the primeval deep.   If the formation of the atmosphere
in its existing conditions was, as already hinted, a result of
the gradual cooling of the earth, then this period must have
been of great length, and the action of the heated waters on
the crust of the globe may have produced thick layers of
detrital matter destined to form the first soils of the succeeding æon.   We know nothing, however, of these primitive
strata, and most of them must have been removed by denud-
ing agencies in succeeding periods, or restored by subterranean heat to the crystalline state.   The events and results of
this day may be summed up as follows :

"At the commencement of the period the earth was enveloped by a misty or vaporous mantle.   In its progress those
relations of air and vapor which cause the separation of the
clouds from the earth by a layer of clear air, and the varied
alternations of sunshine and rain, were established.   At the
close of the period the newly formed atmosphere covered a
universal ocean ; and there was probably a very regular and
uniform condition of the atmospheric currents, and of the
processes of evaporation and condensation."

But while we must affirm that no idea of a solid atmospheric vault can be detected in the Bible, and while we may
also affirm that such an idea would have been altogether foreign to its tone, which invariably refers all things not to secondary machinery, but to the will and fiat of the Supreme, we
must not forget that a most important moral purpose was to
be served by the assertion of the establishment of the atmospheric expanse.   Among all nations the phenomena of the
atmosphere have had important theological and mythological
relations.   The ever-changing and apparently capricious aspects of the atmosphere and its clouds, the terrible effects of
storms, and the balmy influence of sunshine and calm, deeply

impress the minds of simple and superstitious men, and this all the more that in their daily life and expeditions they are constantly subjected to the effects of atmospheric vicissitudes. Hence the greatest gods of all the ancient nations are weather-gods—rulers of the atmospheric heavens—displaying their anger in the thunder-storm and tornado. It is likely that in most cases, as in many barbarous tribes of modern times, these weather-gods were malevolent beings contending against the genial influences of the heavenly Sun-god ; but in nearly every case their supposed practical importance has elevated them, as in the case of the Olympian Zeus, the Scandinavian Thor, and the American Hurakon, to the place of supreme divinity. This was one of the superstitions which the Hebrew monotheism had to overcome. Hence the atmosphere is affirmed to be under Jehovah's law, and all its phenomena are attributed to his power. The value of this as cutting at the root of the most widespread superstitions it is easy to understand, and it has a farther value in teaching that even the apparently unstable and capricious air is a thing established from the first and amenable to the ordinance of God. How difficult it has been to eradicate superstitious views of the atmosphere may be learned from the fact that St. Paul, in writing to the enlightened citizens of Ephesus, could speak of the power which the heathen worshipped as the " Prince of the powers of the air," and it is also evidenced by the abundant notions of this kind which have survived from the Middle Ages among the more ignorant part of the people even in lands called Christian.

While, however, the Bible affirms the atmosphere to be subject to law, it does not carry this into the domain of physical necessity, and affirm with some modern materialistic philosophers that it is useless to pray for rain. It is God who gives

rain from heaven and fruitful seasons, and what he gives
he can withhold. Perhaps no part of our subject can better than this illustrate the rational distinction between a
mere physical fatalism, or a mere superstitious fear of capricious nature, and that belief in a divine Lawgiver which lies
between these extremes. Modern science may smile at the
poor Indian, who in his fear invokes Hurakon or Tlaloc or
the terrible Thunder-bird, and may even despise that nobler
worship of the great Phœnician Sun-god, the source and fount-
ain of all light and life ; against which, though it was the
grandest of all the old idolatries, Elijah waged war to the
death. But may it not equally deride the faith of Elijah
himself, when, after three years of drought, he prayed in the
sight of assembled Israel for rain? It may do so if physical
law amounts to an invariable necessity, and if there is no
supreme Will behind it. But if natural laws are the expression of the divine will, if these laws are multiform and complicated in their relations, and regulate vastly varied causes
interacting with each other, and if the action and welfare of
man come within the scope of these laws, then there is nothing irrational in the supposition that God, without any capricious or miraculous intervention, may have so correlated the
myriad adjustments of his creation as that, while it is his usual
rule that rain falls alike on the evil and on the good, he may
make its descent at particular times and places to depend on
the needs and requests of his own children. In truth the
belief in law is essential to the philosophical conception of
prayer. If the universe were a mere chaos of chances, or
if it were a result of absolute necessity, there would be no
place for intelligent prayer; but if it is under the control of
a Lawgiver, wise and merciful, not a mere manager of material machinery, but a true Father of all, then we can go to

such a being with our requests, not in the belief that we can change his great plans, or that any advantage could result from this if it were possible, but that these plans may be made in his boundless wisdom and love to meet our necessities. There is also in the Bible the farther promise that, if we are truly the children of God, regulating our conduct by his will and enlightened by his spirit, we shall know how to pray for what is in accordance with his divine purpose, and how to receive with gladness whatever he sees fit to give. While, therefore, the Biblical doctrine as to natural law emancipates us from fears of angry storm-demons, it draws us near to a heavenly Father, whose power is above all the tempests of earth, and who, while ruling by law, has regulated all things in conformity with the higher law of love. When God had made the atmosphere, he saw that it was good, and the highest significance is given to this by the consideration that God is love. The position of the Bible is thus the true mean between superstitions at once unhappy and debasing, and a materialistic infidelity that would reduce the universe to a dead, remorseless machine, in which we must struggle for a precarious existence till we are crushed between its wheels.

CHAPTER VIII.

THE DRY LAND AND THE FIRST PLANTS.

"And God said, Let the waters under the heavens be gathered into one place, and let the dry land appear: and it was so. And God called the dry land earth, and the gathering of waters called he seas; and God saw that it was good.

"And God said, Let the earth bring forth the springing herb, the herb bearing seed, and the fruit tree yielding fruit, after its kind, whose seed is in it on the earth: and it was so. And the earth brought forth the tender herb, the herb yielding seed, and the tree bearing fruit whose seed is in it, after its kind; and God saw that it was good."—Genesis i., 10, 11.

THESE are events sufficiently simple and intelligible in their general character. Geology shows us that the emergence of the dry land must have resulted from the elevation of parts of the bed of the ancient universal ocean, and that the agent employed in such changes is the bending and crumpling of the outer crust of the earth, caused by lateral pressure, and operating either in a slow and regular manner or by sudden paroxysms. It farther informs us that the existing continents consist of stratified or bedded masses, more or less inclined, fissured and irregularly elevated, and usually supported by crystalline rocks which have been produced among them, or forced up beneath or through them by internal agencies, and which truly constitute the pillars and foundations of the earth. These elevations, it is true, were successive, and belong to different periods; but the appearance of the first dry land is that intended here.

The elevation of the dry land is more frequently referred to in Scripture than any other cosmological fact ; and while all have been misapprehended, the statements on this subject have been even more unjustly dealt with than others.   In the text, the word "earth" (*arets**) is, by divine sanction, narrowed in meaning to the dry land ; but while some expositors are quite willing to restrict it to this, or even a more limited sense, in the first and second verses of this chapter, almost the only verses in the Bible where the terms of the narrative make such a restriction inadmissible, they are equally ready to understand it as meaning the whole globe in places where the explanatory clause in the verse now under consideration teaches us that we should understand the land only, as distinguished from the sea.   I may quote some of these passages, and note the views they give ; always bearing in mind that, after the intimation here given, we must understand the term "earth" as applying *only to the continents* or *dry land*, unless where the context otherwise fixes the meaning. We may first turn to Psalm civ. :

> "Thou laidst the foundations of the earth,
> That it should never be removed ;
> Thou coveredst it with the deep as with a garment ;
> The waters stood above the mountains ;
> At thy rebuke they fled ;
> At the sound of thy thunder they hasted away ;
> Mountains ascended, valleys descended
> To the place thou hast appointed for them :
> Thou hast appointed them bounds that they may not pass,
> That they return not again to cover the earth."

---

* The word is one of those that pervade both Semitic and Indo-European tongues: Sanscrit, *ahara*; Pehlevi, *arta*; Latin, *terra*; German, *Erde*; Gothic, *airtha*; Scottish, *yird*; English, *earth*.—Gesenius.

The position of these verses in this " the hymn of creation " leaves no doubt that they refer to the events we are now considering.   I have given above the literal reading of the line that refers to the elevation of mountains and subsidence of valleys ; admitting, however, that the grammatical construction gives an air of probability to the rendering in our version, " they go up by the mountains, they go down by the valleys ;" which, on the other hand, is rendered very improbable by the sense.   In whichever sense we understand this line, the picture presented to us by the Psalmist includes the elevation of the mountains and continents, the subsidence of the waters into their depressed basins, and the firm establishment of the dry land on its rocky foundations, the whole accompanied by a feature not noticed in Genesis—the voice of God's thunder —or, in other words, electrical and volcanic explosions.   The following quotations refer to the same subject :

> " Before the mountains were settled,
> Before the hills was I (the Wisdom of God) brought forth ;
> While as yet he had not made the earth,
> Nor the plains, nor the higher parts of the habitable world.
> When he gave the sea his decree
> That the waters should not pass his limits,
> When he determined the foundations of the earth."
> —Proverbs viii., 25.

> " Thou hast established the earth, and it endureth,
> According to thy decrees they continue this day,
> For all are thy servants."
> —Psalm cxix., 90.

> " Who shaketh the earth out of its place,
> And its pillars tremble."
> —Job ix., 6.

> " Where wast thou when I founded the earth ?
> Declare, if thou hast knowledge.

Who hath fixed the proportion thereof, if thou knowest?
Who stretched the line upon it?
Upon what are its foundations settled?
Or who laid its corner-stone,
When the morning stars sang together,
And all the sons of God shouted for joy?
Who shut up the sea with doors
In its bursting forth as from the womb?
When I made the cloud its garment,
And swathed it in thick darkness,
I measured out for it my limit,
And fixed its bars and doors;
And said, Thus far shalt thou come, but no farther,
And here shall thy proud waves be stayed."
　　　　　　　　　　—Job xxxviii., 4.

In these passages the foundation of the earth at first, as well as the shaking of its pillars by the earthquake, are connected with what we usually call natural law—the decree of the Almighty—the unchanging arrangements of an unchangeable Creator, whose "hands formed the dry land."* This is the ultimate cause not only of the elevation of the land, but of all other natural things and processes. The naturalist does not require to be informed that the details, in so far as they are referred to in the above passages, are perfectly in accordance with what we know of the nature and support of continental masses. Geological observation and mathematical calculation have in our day combined their powers to give clear views of the manner in which the fractured strata of the earth are wedged and arched together, and supported by internal igneous masses upheaved from beneath, and subsequently cooled and hardened. A general view of these facts which we have learned from scientific inquiry, the Hebrews

* Psalm xcv.

H 2

gleaned with nearly as much precision from the short account
of the elevation of the land in Genesis, and from the later
comments of their inspired poets. From the same source
our own great poet, Milton, learned these cosmical facts, before the rise of geology, and expressed them in unexception-
able terms ·

> "The mountains huge appear
> Emergent, and their broad bare backs upheave
> Into the clouds, their tops ascend the sky.
> So high as heaved the tumid hills, so low
> Down sunk a hollow bottom, broad and deep,
> Capacious bed of waters."

In further illustration of the opinions of the Scripture
writers respecting the nature of the earth, and the disturbances to which it is liable, I quote the following passages.
The first is from the magnificent description of Jehovah
descending to succor his people amid the terrors of the earthquake, the volcano, and the thunder-storm, in Psalm xviii. :

> "Then shook and trembled the earth,
> The foundations of the hills moved and were shaken,
> Because he was angry.
> Smoke went up from his nostrils,
> Fire from his mouth devoured,
> Coals were kindled by it.
> Then were seen the channels of the waters,
> And the foundations of the world were discovered,
> A tthy rebuke—O Jehovah—
> At the blast of the breath of thy nostrils."

In another place in the Psalms we find volcanic action
thus tersely sketched :

> "He looketh on the earth and it trembleth,
> He toucheth the hills and they smoke."
> —Psalm civ., 32.

Perhaps the most remarkable discourse on this subject in
the whole Bible is that in Job xxviii., in which mining operations are introduced as an illustration of the difficulty of obtaining true wisdom. This passage is interesting both from
its extreme antiquity, and the advancement in knowledge and
practical skill which it indicates. It presents, however, many
difficulties ; and its details have almost entirely lost their
true significance in our common English version :

"Surely there is a vein for silver,
   And a place for the gold which men refine ;
   Iron is taken from the earth,
   And copper is molten from the ore.
   To the end of darkness and to all extremes man searcheth,
   For the stones of darkness and the shadow of death.
   He opens a passage [shaft] from where men dwell,
   Unsupported by the foot, they hang down and swing to and fro.*
   The earth—out of it cometh bread ;
   And beneath, it is overturned as by fire.†
   Its stones are the place of sapphires,
   And it hath lumps‡ of gold.
   The path (thereto) the bird of prey hath not known,
   The vulture's eye hath not seen it.§
   The wild beasts' whelps have not trodden it,
   The lion hath not passed over it.
   Man layeth his hand on the hard rock,
   He turneth up the mountains from their roots,
   He cutteth channels [*adits*] in the rocks,
   His eye seeth every precious thing.

---

* Gesenius.
† Perhaps "changed," metamorphosed, as by fire. Conant has "destroyed."
‡ "Dust" in our version, literally lumps or "nuggets."
§ The vulgar and incorrect idea that the vulture "scents the carrion
from afar," so often reproduced by later poets, has no place in the Bible
poetry. It is the bird's keen eye that enables him to find his prey.

He restraineth the streams from trickling,
And bringeth the hidden thing to light.
But where shall wisdom be found,
And where is the place of understanding?"

This passage, incidentally introduced, gives us a glimpse of the knowledge of the interior of the earth and its products, as it existed in an age probably anterior to that of Moses. It brings before us the repositories of the valuable metals and gems—the mining operations, apparently of some magnitude and difficulty, undertaken in extracting them—and the wonderful structure of the earth itself, green and productive at the surface, rich in precious metals beneath, and deeper still the abode of intense subterranean fires. The only thing wanting to give completeness to the picture is some mention of the fossil remains buried in the earth; and, as the main thought is the eager and successful search for useful minerals, this can hardly be regarded as a defect. The application of all this is finer than almost any thing else in didactic poetry. Man can explore depths of the earth inaccessible to all other creatures, and extract thence treasures of inestimable value; yet, after thus exhausting all the natural riches of the earth, he too often lacks that highest wisdom which alone can fit him for the true ends of his spiritual being. How true is all this, even in our own wonder-working days! A poet of to-day could scarcely say more of subterranean wonders, or say it more truthfully and beautifully; nor could he arrive at a conclusion more pregnant with the highest philosophy than the closing words :

"The fear of the Lord, that is wisdom ;
And to depart from evil is understanding."

The emergence of the dry land is followed by a repetition of the approval of the Creator. "God saw that it was good." To our view that primeval dry land would scarcely have seemed good. It was a world of bare, rocky peaks, and verdureless valleys—here active volcanoes, with their heaps of scoriæ and scarcely cooled lava currents—there vast mud-flats, recently upheaved from the bottom of the waters—nowhere even a blade of grass or a clinging lichen. Yet it was good in the view of its Maker, who could see it in relation to the uses for which he had made it, and as a fit preparatory step to the new wonders he was soon to introduce. Then too, as we are informed in Job xxxviii., "The morning stars sang together, and all the sons of God shouted for joy." We also, when we think of the beautiful variety of the terrestrial surface, the character and composition of its soils, the variety of climate and exposure resulting from its degrees of elevation, the arrangements for the continuance of springs and streams, and many other beneficial provisions connected with the merely mechanical arrangements of the dry land, may well join in the tribute of praise to the All-wise Creator. There is, however, a farther thought suggested by the approval of the great Artificer. In this wondrous progress of creation, it seems as if every thing at first was in its best estate. No succeeding state could parallel the unbroken symmetry of the earth in the fluid and vaporous condition of the "deep." Before the elevation of the land, the atmospheric currents and the deposition of moisture must have been surpassingly regular. The first dry land may have presented crags and peaks and ravines and volcanic cones in a more marvellous and perfect manner than any succeeding continents—even as the dry and barren moon now, in this respect, far surpasses the earth. In the progress of organic

life, geology gives similar indications, in the variety and magnitude of many animal types on their first introduction; so
that this may very possibly be a law of creation.

During the emergence of the first dry land, large quantities
of detrital matter must have been deposited in the waters,
and in part elevated into land. All of these beds would,
probably, be destitute of organic remains ; but if such beds
were formed and still remain, they are probably unknown to
us, for the oldest formations that we know—those of the Eozoic age—contain traces of such remains. It has, indeed,
been suggested that these most ancient organisms are, as it
were, overlooked in the history of creation, or regarded as
equivalent to those shapeless monsters and animals of the
darkness that are referred to in the older Turanian versions
of this story of creation. I doubt very much, however, if this
is a fair interpretation of our ancient record ; but we shall be
in a better position to discuss it when we come to the actual
introduction of animals.

Modern analogy would induce us to believe that the land
was not elevated suddenly; but either by a series of small
paroxysms, as in the case of Chili, or by a gradual and imperceptible movement, as in the case of Sweden—two of the
most remarkable modern instances of elevation of land—accompanied, however, in the case of the last by local subsidence.* In either of these ways the seas and rivers would
have time to smooth the more rugged inequalities, to widen
the ravines into valleys, and to spread out sediment in the
lower grounds ; thus fitting the surface for the habitation of
plants and animals. We must not suppose, however, that the
dry land had any close resemblance to that now existing in

* Lyell's " Principles of Geology."

its form or distribution. Geology amply proves that since the first appearance of dry land, its contour has frequently been changed, and probably also its position. Hence nearly all our present land consists of rocks which have been formed under the waters, long after the period now under consideration, and have been subsequently hardened and elevated ; and since all the existing high mountain ranges are of a comparatively late age, it is probable that this primeval dry land was low, as well as, in the earlier part of the period-at least, of comparatively small extent. It is, however, by no means certain that there may not have been a greater expanse of land toward the close of this period than that which afterwards existed in those older periods of animal life to which the earliest fossiliferous rocks of the geologist carry us back ; since, as already hinted, it seems to be a rule in creation that each new object shall be highly developed of its kind at its first appearance, and since there have been in geological time many great subsidences as well as elevations. Neither must we forget that the oldest land has been subjected throughout geological time to wearing and degrading agencies, and that from its waste the later formations have been mainly derived.

It would be wrong, however, to omit to state that, though we may know at present no remains of the first dry land, we are not ignorant of its general distribution; for the present continents show, in the arrangement of their formations and mountain chains, evidence that they are parts of a plan sketched out from the beginning. It has often been remarked by physical geographers that the great lines of coast and mountain ranges are generally in directions approaching to northeast and southwest, or northwest and southeast, and that where they run in other directions, as in the case of the

south of Europe and Asia, they are much broken by salient
and re-entering angles, formed by lines having these directions.   Professor R. Owen, of Tennessee, and Professor Pierce,
of Harvard College, were, I believe, the first to point out that
these lines are in reality parts of great circles tangent to the
polar circles, and the latter to suggest a theory of their origin, based on the action of solar heat and the seasons on a
cooling earth.   This has been more fully stated by Mr. W.
Lowthian Green in his curious book, " Vestiges of the Molten
Globe."* It would appear that the great circles in question are
in reality at right angles to the line of direction of the attraction of the sun and moon at the period of either solstice, and
when they happen to be in conjunction or opposition at these
periods ; and that such circles would be the lines on which
the thin crust of a cooling globe would be most likely to be
ruptured by its internal tidal-wave.   Whatever the cause of
the phenomenon, it is evident that in the formation of its
surface inequalities the earth has cracked—so to speak—
along two series of great circles tangent to the polar circles ;
and that these, with certain subordinate lines of fracture running north and south and east and west, have determined the
forms of the continents from their origin.

M. Elie de Beaumont, and after him most other geologists,
have attributed the elevation of the continents and the upheaval and plication of mountain chains to the secular refrigeration of the earth, causing its outer shell to become too
capacious for its contracting interior mass, and thus to break
or bend, and to settle toward the centre.   This view would
well accord with the terms in which the elevation of the land
is mentioned throughout the Bible, and especially with the

---

* Stanford, London, 1875.

general progress of the work as we have gleaned it from the
Mosaic narrative; since from the period of the desolate void
and aeriform deep to that now before us secular refrigeration
must have been steadily in progress. Let us also observe
here that the earliest fractures of the crust would determine
the first coast lines, and the first slopes along which sedi-
mentary matter would descend from the land and be deposited in the sea. They would also modify the direction of
the ocean currents. Thus the deposition of new formations
would be directed by these old lines, as would also to some
extent the course of all subsequent fractures and plications.
Thus it happens that the lines of outcrop of the oldest rocks
first raised out of the waters already marked out the forms
of the continents, and that the later formations appear rather
as fillings-up and extensions of the skeleton established by
the first dry land. Farther, the lines of plication first established along the borders of the continents formed resisting
walls along which, in the continued contraction of the earth,
pressure was exerted from the ocean bed, widening and elevating these lines of upheaval, and still farther fixing the
general forms of the continents, and giving variety to their
surfaces. In the progress of geological time there have also
been successive depressions and re-elevations of the continental plateaus, subjecting them alternately to the wearing
and disintegrating action of the atmosphere and its waters,
and to the influence of waves and ocean currents, and especially to that of the deep-seated polar currents which have
throughout geological ages been loading the submerged
areas of the earth's surface with the products of the waste
caused by frost and ice in the polar regions. These causes
again have been progressively increasing the oblateness of the
earth's figure, and, along with the slackening of its rotation,

preparing the way for those periodical collapses in the equatorial and temperate regions which form the boundaries of some of our most important geological periods.*   Throughout all these changes the great general plan of the continents, first sketched out when the "foundations of the earth" were laid, before Eozoic time, was being elaborated.

The same creative period that witnessed the first appearance of dry land saw it also clothed with vegetation; and it is quite likely that this is intended to teach that no time was lost in clothing the earth with plants—that the first emerging portions received their vegetable tenants as they became fitted for them—and that each additional region, as it rose above the surface of the waters, in like manner received the species of plants for which it was adapted.   What was the nature of this earliest vegetation?   The sacred writer specifies three descriptions of plants as included in it; and, by considering the terms which he uses, some information on this subject may be gained.

Deshé, translated "grass" in our version, is derived from a verb signifying to spring up or bud forth; the same verb, indeed, used in this verse to denote "bringing forth," literally causing to spring up.   Its radical meaning is, therefore, vegetation in the act of sprouting or springing forth; or, as connected with this, young and delicate herbage.   Thus, in Job xxxviii., "To satisfy the desolate and waste ground, and to cause the bud of the *young herbage* to spring forth."   Here the reference is, no doubt, to the bulbous and tuberous rooted plants of the desert plains, which, fading away in the summer drought, burst forth with magical rapidity on the setting-in

---

* In further explanation of these general geological changes, see "The Story of the Earth and Man," by the author.

of rain. The following passages are similar: Psalm xxiii.,
"He maketh me to lie down in green pastures" (literally,
young or *tender herbage*); Deuteronomy xxiii., "Small rain
upon the *tender herb;*" Isaiah xxxvii., "*Grass* on the house-
tops." The word is also used for herbage such as can be
eaten by cattle or cut down for fodder, though even in these
cases the idea of young and tender herbage is evidently included ; "Fat as a heifer at *grass*" (Jer. xiv.)—that is, feeding
on young succulent grass, not that which is dry and parched.
"Cut down as the grass, or wither as the green herb," like
the soft, tender grass, soon cut down and quickly withering.
With respect to the use of the word in this place, I may remark : 1. It is not here correctly translated by the word "grass;"
for grass bears seed, and is, consequently, a member of the
second class of plants mentioned.   Even if we set aside all
idea of inspiration, it is obviously impossible that any one
living among a pastoral or agricultural people could have
been ignorant of this fact.   2. It can scarcely be a general
term, including all plants when in a young or tender state.
The idea of their springing up is included in the verb, and
this was but a very temporary condition.   Besides, this word
does not appear to be employed for the young state of shrubs
or trees.   3. We thus appear to be shut up to the conclusion
that *deshé* here means those plants, mostly small and herbaceous, which bear no proper seeds;* in other words, the
Cryptogamia—as fungi, mosses, lichens, ferns, etc.   The remaining words are translated with sufficient accuracy in our
version.   They denote seed-bearing or phœnogamous herbs
and trees.   The special mention of the fructification of
plants is probably intended not only for distinction, but also

---

* "Tenera herba, sine semine saltem conspicuo."—Rosenmüller, " Scho-
lia."

to indicate the new power of organic reproduction now first
introduced on the surface of our planet, and to mark its difference from the creative act itself. That this new and wondrous phenomenon should be so stated is thus in strict scientific propriety, and it is precisely the point that would be
seized by an intelligent spectator of the visions of creation,
who had previously witnessed only the accretion and disintegration of mineral substances, and to whom this marvellous
power of organic reproduction would be in every respect a
new creation.

The arrangement of plants in the three great classes of
cryptogams, seed-bearing herbs, and fruit-bearing trees differs in one important point—viz., the separation of herbaceous
plants from trees—from modern botanical classification. It
is, however, sufficiently natural for the purposes of a general
description like this, and perhaps gives more precise ideas of
the meaning intended than any other arrangement equally
concise and popular.   It is also probable that the object of
the writer was not so much a natural-history classification as
an account of the *order* of creation, and that he wishes to affirm that the introduction of these three classes of plants on
the earth corresponded with the order here stated.   This
view renders it unnecessary to vindicate the accuracy of the
arrangement on botanical grounds, since the historical order
was evidently better suited to the purpose in view, and in so
far as the earlier appearance of cryptogamous plants is concerned, it is in strict accordance with geological fact.

A very important truth is contained in the expression
"after its kind"—that is, after its *species;* for the Hebrew
"*min,*" used here, has strictly this sense, and, like the Greek
idea and the Latin *species*, conveys the notion of form as well
as that of kind.   It is used to denote species of animals,

in Leviticus i., 14, and in Deuteronomy xiv., 15. We are taught by this statement that plants were created each kind by itself, and that creation was not a sort of slump-work to be perfected by the operation of a law of development, as fancied by some modern speculators. In this assertion of the distinctness of species, and the production of each as a distinct part of the creative plan, revelation tallies perfectly with the conclusions of natural science, which lead us to believe that each species, as observed by us, is permanently reproductive, variable within narrow limits, and incapable of permanent intermixture with other species; and though hypotheses of modification by descent, and of the production of new species by such modification, may be formed, they are not in accordance with experience, and are still among the unproved speculations which haunt the outskirts of true science. We shall be better prepared, however, to weigh the relations of such hypotheses to our revelation of origins when we shall have reached the period of the introduction of animal life.

Some additional facts contained in the recapitulation of the creative work in Chapter II. may very properly be considered here, as they seem to refer to the climatal conditions of the earth during the growth of the most ancient vegetation, and before the final adjustment of the astronomical relations of the earth on the fourth day. "And every shrub of the land before it was on the earth, and every herb of the land before it sprung up. For the Lord God had not caused it to rain on the earth, and there was not a man to till the ground ; but a mist ascended from the earth and watered the whole surface of the ground." This has been supposed to be a description of the state of the earth during the whole period anterior to the fall of man. There is, however, no Scripture evidence of this ; and geology informs us that rain fell as at

present far back in the Palæozoic period, countless ages before the creation of man or the existing animals.   Although,
however, such a condition of the earth as that stated in these
verses has not been known in any geological period, yet it is
not inconceivable, but in reality corresponds with the other
conditions of nature likely to have prevailed on the third day,
as described in Genesis. ' The land of this period, we may
suppose, was not very extensive nor very elevated.   Hence
the temperature would be uniform and the air moist.   The
luminous and calorific matter connected with the sun still
occupied a large space, and therefore diffused heat and light
more uniformly than at present.   The internal heat of the
earth may still have produced an effect in warming the oceanic waters.   The combined operation of these causes, of
which we, perhaps, have some traces as late as the Carboniferous period, might well produce a state of things in which
the earth was watered, not by showers of rain, but by the
gentle and continued precipitation of finely divided moisture,
in the manner now observed in those 'climates in which vegetation is nourished for a considerable part of the year by
nocturnal mists and copious dews.   The atmosphere, in short,
as yet partook in some slight degree of the same moist and
misty character which prevailed before the " establishment of
the clouds above"—the airy firmament of the second day.
The introduction of these explanatory particulars by the sacred
historian furnishes an additional argument for the theory of
long periods.   That vegetation should exist for two or three
natural days without rain or the irrigation which is given in
culture, was, as already stated, a circumstance altogether unworthy of notice ; but the growth during a long period of a
varied and highly organized flora, without this advantage, and
by the aid of a special natural provision afterward discontin-

ued, was in all respects so remarkable and so highly illustrative of the expedients of the divine wisdom that it deserved a prominent place.

It is evident that the words of the inspired writer include plants belonging to all the great subdivisions of the vegetable kingdom. This earliest vegetation was not rude or incomplete, or restricted to the lower forms of life. It was not even, like that of the coal period, solely or mainly cryptogamous or gymnospermous. It included trees bearing fruit, as well as lichens and mosses, and it received the same stamp of approbation bestowed on other portions of the work—" it was good." We have a good right to assume that its excellence had reference not only to its own period, but to subsequent conditions of the earth. Vegetation is the great assimilating power, the converter of inorganic into organic matter suitable for the sustenance of animals. In like manner the lower tribes of plants prepare the way for the higher. We should therefore have expected *à priori* that vegetation would have clothed the earth before the creation of animals, and a sufficient time before it to allow soils to be accumulated, and surplus stores of organic matter to be prepared in advance : this consideration alone would also induce us to assign a considerable duration to the third day. After the elevation of land, and the draining off from it of the saline matter with which it would be saturated, a process often very tedious, especially in low tracts of ground, the soil would still consist only of mineral matter, and must have been for a long period occupied by plants suited to this condition of things, in order that sufficient organic matter might be accumulated for the growth of a more varied vegetation ; a consideration which perhaps illustrates the order of the plants in the narrative.

It may be objected to the above views that, however ac-

-cordant with chemical and physiological probabilities, they do not harmonize with the facts of geology ; since the earliest fossiliferous formations contain almost exclusively the remains of animals, which must therefore have preceded, or at least been coeval with, the earliest forms of terrestrial vegetation. This objection is founded on well-ascertained facts, but facts which may have no connection with the third day of creation when regarded as a long period. The oldest geological formations are of marine origin, and contain remains of marine animals, with those of plants supposed to be allied to the existing algæ or sea-weeds. Geology can not, however, assure us either that no land plants existed contemporaneously with these earliest animals, or that no land flora preceded them. These oldest fossiliferous rocks may mark the commencement of animal life, but they testify nothing as to the existence or non-existence of a previous period of vegetation alone. Farther, the rocks which contain the oldest remains of life exist as far as yet known in a condition so highly metamorphic as almost to preclude the possibility of their containing any distinguishable vegetable fossils ; yet they contain vast deposits of carbon in the form of graphite, and if this, like more modern coaly matter, was accumulated by vegetable growth, it must indicate an exuberance of plants in these earliest geological periods, but of plants as yet altogether unknown to us. It is possible, therefore, that in these Eozoic rocks we may have remnants of the formations of the third Mosaic day ; and if we should ever be so fortunate as to find any portion of them containing vegetable fossils, and these of species differing from any hitherto known, either in a fossil state or recent, and rising higher, in elevation and complexity of type, than the flora of the succeeding Silurian and Carboniferous eras, we may then suppose that we have penetrated to the monuments of this third creative æon. The only other alternative by which these verses can be reconciled with geology is that adopted by the late Hugh Miller, who supposes that the plants of the third day are those of the Carboniferous period; but, besides the apparent anachronism involved in this, we now know that the coal flora consisted mainly of cryptogams allied to ferns and club-mosses, and of gymnosperms allied to the pines and cycads, the higher orders of plants being almost entirely wanting. For these reasons we are shut up to the conclusion that this flora of the third day must have its place before the Palæozoic period of geology.

To those who are familiar with the vast lapse of time required by the geological history of the earth, it may be startling to ascribe the whole of it to three or four of the creative days. If, however, it be admitted that these days were periods of unknown duration, no reason remains for limiting their length any farther than the facts of the case require. If in the strata of the earth which are accessible to us we can detect the evidence of its existence for myriads of years, why may not its Creator be able to carry our view back for myriads more. It may be humbling to our pride of knowledge, but it is not on any scientific ground improbable, that the oldest animal remains known to geology belong to the middle period of the earth's history, and were preceded by an enormous lapse of ages in which the earth was being prepared for animal existence, but of which no records remain, except those contained in the inspired history.

It would be quite unphilosophical for geology to affirm either that animal life must always have existed, or that its earliest animals are necessarily the earliest organic beings. To use, with a slight modification, the words of an able think-

I

er on these subjects,* " For ages the prejudice prevailed that
the historical period, or that which is coeval with the life of
man, exhausted the whole history of the globe. Geologists
removed that prejudice," but must not substitute "another in
its place, viz., that geological time is coeval with the globe itself, or that organic life always existed on its surface."

A second doubt as to the existence of this primitive flora
may be based on the statement that it included the highest
forms of plants. Had it consisted only of low and imperfect
vegetables, there might have been much less difficulty in admitting its probability. Farther, we find that even in the
Carboniferous period scarcely any plants of the higher orders
flourished, and there was a preponderance of the lower forms
of the vegetable kingdom. We have, however, in geological
chronology, many illustrations of the fact that the progress
of improvement has not been continuous or uninterrupted,
and that the preservation of the flora and fauna of many geological periods has been very imperfect. Hence the occurrence in one particular stratum or group of strata of few or
low representatives of animal and vegetable life affords no
proof that a better state of things may not have existed previously. We also find, in the case of animals, that each tribe
attained to its highest development at the time when, in the
progress of creation, it occupied the summit of the scale of
life. Analogy would thus lead us to believe that when plants
alone existed, they may have assumed nobler forms than any
now existing, or that tribes now represented by few and humble species may at that time have been so great in numbers
and development as to fill all the offices of our present complicated flora, as well as, perhaps, some of those now occupied

---

* Haughton, Address to the Geological Society, Dublin.

by animals. We have this principle exemplified in the Car-boniferous flora, by the magnitude of its arborescent club-mosses, and the vast variety of its gymnosperms. For this reason we may anticipate that if any remains of this early plant-creation should be disinterred, they will prove to be among the most wonderful and interesting geological relics ever discovered, and will enlarge our views of the compass and capabilities of the vegetable kingdom, and especially of its lower forms.

A farther objection is the uselessness of the existence of plants for a long period, without any animals to subsist on or enjoy them, and even without forming any accumulation of fossil fuel or other products useful to man. The only direct answer to this has already been given. The previous exist-ence of plants may have been, and probably was, essential to the comfort and subsistence of the animals afterwards intro-duced. Independently of this, however, we have an analo-gous case in the geological history of animals, which prevents this fact from standing alone. Why was the earth tenanted so long by the inferior races of animals, and why were so much skill and contrivance expended on their structures, and even on their external ornament, when there was no intelligent mind on earth to appreciate their beauties. Even in the pres-ent world we may as well ask why the uninhabited islands of the ocean are found to be replete with luxuriant vegetable life, why God causes it to rain in the desert where human foot never treads, or why he clothes with a marvellous exuberance of beautiful animal and plant forms the depths of the sea. We can but say that these things seemed and seem good to the Creator, and may serve uses unknown to us ; and this is pre-cisely what we must be content to say respecting the plant-creation of the Eozoic period.

Some writers* on this subject have suggested that the cosmical use of this plant-creation was the abstraction from the atmosphere of an excess of carbonic acid unfavorable to the animal life subsequently to be introduced. This use it may have served, and when its effects had been gradually lost through metamorphism and decay, that second great withdrawal of carbon which took place in the Carboniferous period may have been rendered necessary. The reasons afforded by natural history for supposing that plants preceded animals are thus stated by Professor Dana:

"The proof from science of the existence of plants before animals is inferential, and still may be deemed satisfactory. Distinct fossils have not been found, all that ever existed in the azoic† rocks having been obliterated. The arguments in the affirmative are as follows:

" 1. The existence of limestone rocks among the other beds, similar limestones in later ages having been of organic origin; also the occurrence of carbon in the shape of graphite, graphite being, in known cases in rocks, a result of the alteration of the carbon of plants.

" 2. The fact that the cooling earth would have been fitted for vegetable life for a long age before animals could have existed; the principle being exemplified everywhere that the earth was occupied at each period with the highest kinds of life the conditions allowed.

" 3. The fact that vegetation subserved an important pur-

---

* See McDonald, "Creation and the Fall." Professor Guyot, I believe, deserves the credit of having first mentioned, on the American side of the Atlantic, the doctrine respecting the introduction of plants advocated in this chapter.

† "Eozoic" of this work. Professor Dana in the latest edition of his Manual uses the name "Archaean."

pose in the coal-period in ridding the atmosphere of carbonic acid for the subsequent introduction of land animals, suggests a valid reason for believing that the same great purpose, the true purpose of vegetation, was effected through the ocean before the *waters* were fitted for animal life.

" 4. Vegetation being directly or mediately the food of animals, it must have had a previous existence. The latter part of the azoic age in geology we therefore regard as the age when the plant kingdom was instituted, the latter half of the third day in Genesis. However short or long the epoch, it was one of the great steps of progress."

In concluding the examination of the work of the third day, I must again remind the reader that, on the theory of long creative periods, the words under consideration must refer to the first introduction of vegetation, in forms that have long since ceased to exist. Geology informs us that in the period of which it is cognizant the vegetation of the earth has been several times renewed, and that no plants of the older and middle geological periods now exist. We may therefore rest assured that the vegetable species, and probably also many of the generic and family forms of the vegetation of the third day, have long since perished, and been replaced by others suited to the changed condition of the earth. It is indeed probable that during the third and fourth days themselves there might be many removals and renewals of the terrestrial flora, so that perhaps every species created at the commencement of the introduction of plants may have been extinct before the close of the period. Nevertheless it was marked by the introduction of vegetation, which in one or another set of forms has ever since clothed the earth.

At the commencement of the third day the earth was still covered by the waters. As time advanced islands and mountain-peaks arose from the ocean, vomiting forth the molten and igneous materials of the interior of the earth's crust. Plains and valleys were then spread around, rivers traced out their beds, and the ocean was limited by coasts and divided by far-stretching continents. At the command of the Creator plants sprung from the soil—the earliest of organized structures—at first probably few and small, and fitted to contend against the disadvantages of soils impregnated with saline particles and destitute of organic matter; but as the day advanced increasing in number, magnitude, and elevation, until at length the earth was clothed with a luxuriant and varied vegetation, worthy the approval of the Creator, and the admiring song of the angelic "sons of God."

CHAPTER IX.

LUMINARIES.

"And God said, Let there be luminaries in the expanse of heaven, to divide the day from the night ; and let them be for signs and for seasons, and for days and for years. And let them be for luminaries in the expanse of heaven, to give light on the earth : and it was so.

"And God made two great luminaries, the greater luminary to preside over the day, the lesser luminary to preside over the night. He made the stars also. And God placed them in the expanse of heaven to give light on the earth, and to preside over the day and over the night, and to separate the light from the darkness : and God saw that it was good. And the evening and the morning were the fourth day."—Genesis i., 14-19.

AFTER so long a sojourn on the earth, we are in these verses again carried to the heavens. Every scientific reader is struck with the position of this remarkable statement, interrupting as it does the progress of the organic creation, and constituting a break in the midst of the terrestrial history which is the immediate subject of the narrative ; thus, in effect, as has often been remarked, dividing the creative week into two portions. Why was the completion of the heavenly bodies so long delayed? Why were light and vegetation introduced previously? If we can not fully answer these questions, we may at least suppose that the position of these verses is not accidental, though certainly not that which would have been chosen for its own sake by any fabricator of systems ancient or modern. Let us inquire, however, what are the precise terms of the record.

1. The word here used to denote the objects produced clearly distinguishes them from the product of the first day's creation. Then God said, "Let *light* be;" he now says, "Let *luminaries* or light-bearers be." We have already seen that the light of the first day may have emanated from an extended luminous mass, at first occupying the whole extent of the solar system, and more or less attached to the several planetary bodies, and afterwards concentrated within the earth's orbit. The verses now under consideration inform us that the process of concentration was now complete, that our great central luminary had attained to its perfect state. This process of concentration may have been proceeding during the whole of the intervening time, or it may have been completed at once by some more rapid process of the nature of a direct interposition of creative power.

2. The division of light from darkness is expressed by the same terms, and is of the same nature with that on the first day. This separation was now produced in its full extent by the perfect condensation of the luminiferous matters around the sun.

3. The heavenly bodies are said to be intended for *signs*— that is, for marks or indications—either of the seasons, days, and years afterwards mentioned, or of the majesty and power of the true God, as the Creator of objects so grand and elevated as to become to the ignorant heathen objects of idolatrous worship; or perhaps of the earthly events they are supposed to influence. The arrangements now perfected for the first time enabled natural days, seasons, and years to have their limits accurately marked. Previously to this period there had been no distinctly marked seasons, and consequently no natural separation of years, nor were the limits of days at all accurately defined.

4. The terms *expanse* and *heaven*, previously applied to the atmosphere, are here combined to denote the more distant starry and planetary heavens. There is no ambiguity involved in this, since the writer must have well known that no one could so far mistake as to suppose that the heavenly bodies are placed in that atmospheric expanse which supports the clouds.

5. The luminaries were *made* or appointed to their office on the fourth day. They are not said to have been created, being included in the creation of the beginning. They were now completed, and fully fitted for their work. An important part of this fitting seems to have been the setting or placing them in the heavens, conveying to us the impression that the mutual relations and regular motions of the heavenly bodies were now for the first time perfected.

6. The stars are introduced in a parenthetical manner, which leaves it doubtful whether we are merely informed in general terms that they are works of God, as well as those heavenly bodies which are of more importance to us, or that they were arranged as heavenly luminaries useful to our earth on the fourth day. The term includes the fixed stars, and it is by no means probable that these were in any way affected by the work referred to the fourth day, any farther than their appearance from our earth is concerned. This view is confirmed by the language of the 104th Psalm, which in this part of the work mentions the sun and moon alone, without the fixed stars or planets.

It is evident that the changes referred to this period related to the whole solar system, and resulted in the completion of that system in the form which it now bears, or at least in the final adjustment of the motions and relations of the earth; and we have reason to believe that the condensation of the luminous envelope around the sun was one of the most important of these changes. On the hypothesis of La Place, already referred to as most in accordance with the earlier stages of the work, there seems to be no especial reason why the completion of the process of elaboration of the sun and planets should be accelerated at this particular stage. We can easily understand, however, that those closing steps which brought the solar system into a state of permanent and final equilibrium would form a marked epoch in the work; and we can also understand that now, on the eve of the introduction of animal life, there is a certain propriety in the representation of the Creator interfering to close up the merely inorganic part of his great work, and bring this department at least to its final perfection. The fourth day, then, in geological language, marks *the complete introduction of " existing causes " in inorganic nature,* and we henceforth find no more creative interference, except in the domain of organization. This accords admirably with the deductions of modern geology, and especially with that great principle so well expounded by Sir Charles Lyell, and which forms the true basis of modern geological reasonings—that we should seek in existing causes of change for the explanation of the appearances of the rocks of the earth's crust. Geology probably carries us back to the introduction of animal life; and shows us that since that time land, sea, and atmosphere, summer and winter, day and night—all the great inorganic conditions affecting animal life—have existed as at present, and have been subject to modifications the same in kind with those which they now experience, though perhaps different in degree. In this ancient record we find in like manner that the period immediately preceding the creation of animals witnessed the completion of all the great general arrangements on which these phenomena depend. The Bible, therefore, and science agree in the truth that existing causes have been in full force since the creation of animals; and that since that period the exercise of creative power has been limited to the organic world. This has a curious bearing, not often thought of, on modern theories of evolution as compared with the teaching of the Bible. In one important sense, absolute creation, in so far as the inorganic universe is concerned, is in our Mosaic narrative limited to the production of matter and force at first. All else is called making, forming, or appointing. Thus the production of all the arrangements of the waters, the atmosphere, the earth, and the heavens, in the work of the first four days, and even the introduction of plants, may be correctly termed an evolution or development from preformed materials, with the single exception that the reproductive power and specific diversities of plants are recognized as entirely new facts. Creation is properly resumed when animal life is introduced. Hence, in so far as a comparison with the terms of Genesis is concerned, hypotheses as to the evolution of animal life from inorganic matter are in a different position from hypotheses as to the previous evolution of the parts of inorganic nature; and still more so from statements as to the progress of inorganic nature subsequent to the introduction of animals; since within that period, which really includes the whole of geological time, absolutely no creation whatever in the domain of inanimate nature is affirmed in the Biblical record to have taken place. On the contrary, all the arrangements of inorganic nature are represented as finally completed before the creation of animals.

The obliquity of the earth's axis, which gives us the changes of the seasons, is apparently included in the arrangements

of the fourth creative day. The cause of this obliquity, and
the time when it may have attained to its present amount,
have been fertile themes of discussion. It is clear, however,
that if this obliquity was established, as appears to be stated
here, before the introduction of animal life, it can have no
bearing ,on the changes of climate of which we have evidence in geological time since the dawn of animal life, unless, indeed, it is capable of greater variation than astronomers admit; and the same remark applies to supposed
changes in the position of the poles themselves. There is,
however, nothing in this record to oppose the idea of any
secular changes in these arrangements under the laws appointed in the fourth creative period.

The record relating to the fourth day is silent respecting
the mundane history of the period; and geology gives no
very certain information concerning it. If, however, we assume that any of the Eozoic or pre-eozoic rocks are deposits
of this or the preceding period, we may infer from the disturbances and alteration which these have suffered, prior to
the deposition of the Cambrian and Silurian, that during or
toward the close of this day the crust of the earth was affected by great movements. There is another consideration
also leading to important conclusions in relation to this period. In the earliest fossiliferous rocks there seems to be
good evidence that the dry land contemporary with the seas
in which they were formed was of very small extent. Now,
since on the third day a very plentiful and highly developed
vegetation was produced, we may infer that during that period the extent of dry land was considerable, and was probably
gradually increasing. If, then, the Cambrian and Silurian
systems, so rich in marine organic remains, belong to the
commencement of the fifth day, we must conclude that dur-

ing the fourth much of the land previously existing had been again submerged. In other words, during the third day the extent of terrestrial surface was increasing, on the fourth day it diminished, and on the fifth it again increased, and probably has on the whole continued to increase up to the present time. One most important geological consequence of this is that the marine animals of the fifth day probably commenced their existence on sea bottoms which were the old soil surfaces of submerged continents previously clothed with vegetation, and which consequently contained much organic matter fitted to form a basis of support for the newly created animals.

I shall close my remarks on the fourth day by a few quotations from those passages of Scripture which refer to the objects of this day's work. I have already referred to that beautiful passage in Deuteronomy where the Israelites are warned against the crime of worshipping those heavenly bodies which the Lord God hath "divided to every nation under the whole heaven." In the book of Job also we find that the heavenly bodies were in his day regarded as signal manifestations of the power of God, and that several of the principal constellations had received names:

> "He commandeth the sun, and it shineth not;
> He sealeth up the stars;*
> He alone spreadeth out the heavens,
> And walketh on the high waves of the sea;†

---

* This may refer to an eclipse, but from the character of the preceding verses more probably to the obscurity of a tempest. It is remarkable that eclipses, which so much strike the minds of men and affect them with superstitious awe, are not distinctly mentioned in the Old Testament, though referred to in the prophetical parts of the New Testament.

† Perhaps rather the high places of the waters, referring to the atmospheric waters.

He maketh Arcturus, Orion,
The Pleiades, and the hidden chambers of the south;
Who doeth great things past finding out;
Yea, marvellous things beyond number."
—Job ix., 9.

"Canst thou tighten the bonds of the Pleiades,*
Or loose the bands of Orion?
Canst thou bring forth the Mazzaroth in their season,
Or lead forth Arcturus and its sons?
Knowest thou the laws of the heavens,
Or hast thou appointed their dominion over the earth?"
—Job xxxviii., 31.

I may merely remark on these passages that the chambers of the south are supposed to be those parts of the southern heavens invisible in the latitude in which Job resided. The bonds of Pleiades and of Orion probably refer to the apparently close union of the stars of the former group, and the wide separation of those of the latter; a difference which, to the thoughtful observer of the heavens, is more striking than most instances of that irregular grouping of the stars which still forms a question in astronomy, from the uncertainty whether it is real, or only an optical deception arising from stars at different distances coming nearly into a line with each other. I have seen in some recent astronomical work this very instance of the Pleiades and Orion taken as a marked illustration of this problematical fact in astronomy. *Mas-*

---

* The rendering "sweet influences" in our version may be correct, but the weight of argument appears to favor the view of Gesenius that the close bond of union between the stars of this group is referred to. I think it is Herder who well unites both views, the Pleiades being bound together in a sisterly union, and also ushering in the spring by their appearance above the horizon. Conant applies the whole to the seasons, the bands of Orion being in this view those of winter.

zaroth are supposed by modern expositors to be the signs of the Zodiac.

On the whole, the Hebrew books give us little information as to the astronomical theories of the time when they were written. They are entirely non-committal as to the nature of the connections and revolutions of the heavenly bodies ; and indeed regard these as matters in their time beyond the grasp of the human mind, though well known to the Creator and regulated by his laws. From other sources we have facts leading to the belief that even in the time of Moses, and certainly in that of the later Biblical writers, there was not a little practical astronomy in the East, and some good theory. The Hindoo astronomy professes to have observations from 3000 B.C., and the arguments of Baily and others, founded on internal evidence, give some color of truth to the claim. The Chaldeans at a very early period had ascertained the principal circles of the sphere, the position of the poles, and the nature of the apparent motions of the heavens as the results of revolution on an inclined axis. The Egyptian astronomy we know mainly from what the Greeks borrowed from it. Thales, 640 B.C., taught that the moon is lighted by the sun, and that the earth is spherical, and the position of its five zones. Pythagoras, 580 B.C., knew, in addition to the sphericity of the earth, the obliquity of the ecliptic, the identity of the evening and morning star, and that the earth revolves round the sun. This Greek astronomy appears immediately after the opening of Egypt to the Greeks ; and both these philosophers studied in that country. Such knowledge, and more of the same character, may therefore have existed in Egypt at a much earlier period.

The Psalms abound in beautiful references to the creation of the fourth day :

" When I consider the heavens, the work of thy fingers,
The moon and the stars, which thou hast ordained ;
What is man, that thou art mindful of him ?
Or the son of man, that thou visitest him ?"
— P alm viii.

" Who telleth the number of the stars,
Who calleth them all by their names.
Great is our Lord, and of great praise ;
His understanding is infinite.
The Lord lifteth up the meek ;
He casteth the wicked to the ground."
—Psalm cxlvii.

" The heavens declare the glory of God,
The firmament showeth his handiwork ;
Day unto day uttereth speech,
Night unto night showeth knowledge.
They have no speech nor language,
Their voice is not heard ;
Yet their line is gone out to all the earth,
And their words to the end of the world.
In them hath he set a pavilion for the sun,
Which is as a bridegroom coming out of his chamber,
And rejoiceth as a strong man to run a race.
Its going forth is from the end of the heavens,
And its circuit unto the end of them.
And there is nothing hid from the heat thereof."
—Psalm xix.

These are excellent illustrations of the truth of the Scripture mode of treating natural objects, in connection with their Maker. It is but a barren and fruitless philosophy which sees the work and not its author—a narrow piety which loves God but despises his works. The Bible holds forth the golden mean between these extremes, in a strain of lofty poetry and acute perception of the great and beautiful, whether seen in the Creator or reflected from his works.

The work of this day opens up a wide field for astronomical illustration, more especially in relation to the wisdom and benevolence of the Creator as displayed in the heavens ; but it would be foreign to our present purpose to enter into these.

It may be well, however, to think for a moment of the importance of the facts suggested by the writer of Genesis in mentioning the use of the heavenly bodies as signs of time. To what extent civilization or even the continued existence of man as an intelligent being would have been possible without the marks of subdivision of time given by the great astronomical clock of the universe, it is almost impossible for us to imagine. Without such marks of time, in any case, the whole fabric of human culture must have been different from what it is. Farther, in connection with this, it is a grand thought of our early revelation that all these heavenly bodies, however magnificent, and however they might seem to the heathen to be objects of worship, are but marks on God's clock, parts of a mere machine which keeps time for us, and is therefore our servant, as the children of the great Artificer, and not our ruler. The idea has been termed an astrological one ; but astrology as a means of divination has no place in the record. The heavenly bodies are under the law of the Creator, and their function relatively to us is to give light and to give time. Astrological divination is an outgrowth of the Sabæan idolatry, and held in abomination by the monotheistic author of Genesis. His object may be summed up in the following general statements :

1. The heavenly hosts and their arrangements are the work of Jehovah, and are regulated wholly by his laws or ordinances ; a striking illustration of the recognition by the Hebrew writer both of creative interference, and that stable,

natural law which too often withdraws the mind of the philosopher from the ideas of creation and of providence.

2. The heavenly bodies have a relation to the earth—are parts of the same plan, and, whatever other uses they were made to serve, were made for the benefit of man.

3. The general physical arrangements of the solar system were perfected before the introduction of animals on our planet.

CHAPTER X.

THE LOWER ANIMALS.

"And God said, Let the waters swarm with swarming living creatures, and let birds fly on the surface of the expanse of heaven. And God created great reptiles, and every living moving thing, which the waters brought forth abundantly, after their kind, and every bird after its kind; and God saw that it was good.

"And God blessed them, saying, Be fruitful and multiply, and fill the waters of the seas, and let the flying creatures multiply in the earth. And the evening and the morning were the fifth day."—Genesis i., 20–23.

In these words, so full of busy, active, thronging life, we now enter on that part of the earth's history which has been most fully elucidated by geology, and we have thus an additional reason for carefully weighing the terms of the narrative, which here, as in other places, contain large and important truths couched in language of the simplest character.

1. In accordance with the views now entertained by the best lexicographers, the word translated in our version "creeping things" has been rendered "prolific or swarming creatures." The Hebrew is *Sheretz*, a noun derived from the verb used in this verse to denote bringing forth abundantly. It is loosely translated in the Septuagint *Erpeta*, reptiles; and this view our English translators appear to have adopted, without, perhaps, any very clear notions of the creatures intended. The manner in which it is used in other passages places its true meaning beyond doubt. I select as illustrations of the

most apposite character those verses in Leviticus in which
clean and unclean animals are specified, and in which we
have a right to expect the most precise zoological nomenclature that the Hebrew can afford.   In Leviticus xi., 20–23,
insects are defined to be *flying sheretzim*, and in verse 29, etc.,
under the designation *"sheretzim of the land,"* we have animals named in our version the weasel, mouse, tortoise, ferret,
chameleon, lizard, snail, and mole.   The first of these animals is believed to have been a burrowing creature, perhaps
a mole; the second, from the meaning of its name, "ravager
of fields," is thought to have been a mouse.   Some doubt,
however, attends both of these identifications, but it appears
certain that the remaining six species are small reptiles, principally lizards.   We learn, therefore, that the smaller reptiles,
and *perhaps* also a few small mammals, are *sheretzim*.   In
verses 41 and 42 we are introduced to other tribes.   "And
every *sheretz* that swarmeth on the earth shall be an abomination unto you; it shall not be eaten; whatsoever goeth upon
the belly (serpents, worms, snails, etc.), and whatsoever hath
more feet (than four) (insects, arachnidans, myriapods).   In
verses 9 and 10 of the same chapter we have an enumeration
of the *sheretzim* of the waters : "Whatsoever hath fins and
scales in the waters, in the seas and in the rivers, them shall
ye eat.   And all that have not fins and scales in the seas and
the rivers, of all that swarm in the waters (all the *sheretzim*
of the waters), they shall be an abomination unto you."   Here
the general term *sheretz* includes all the fishes and the invertebrate animals of the waters.   From the whole of the above
passages we learn that this is a general term for all the invertebrate animals and the two lower classes of vertebrates, or,
in other words, for the whole animal kingdom except the
mammalia and birds.   To all these creatures the name is

particularly appropriate, all of them being oviparous or ovo-viviparous, and consequently producing great numbers of young and multiplying very rapidly. The only other creatures which can be included under the term are the two doubtful species of small mammals already mentioned. Nothing can be more fair and obvious than this explanation of the term, based both on etymology and on the precise nomenclature of the ceremonial law. We conclude, therefore, that the prolific animals of the fifth day's creation belonged to the three Cuvierian sub-kingdoms of the Radiata, Articulata, and Mollusca, and to the classes of Fish and Reptiles among the vertebrata.

2. One peculiar group of *sheretzim* is especially distinguished by name — the *tanninim*, or "great whales" of our version. It would be amusing, had we time, to notice the variety of conjectures to which this word has given rise, and the perplexities of commentators in reference to it. In our version and the Septuagint it is usually rendered dragon ; but in this place the seventy have thought proper to put *Ketos* (whale), and our translators have followed them. Subsequent translators and commentators have laid under contribution all sorts of marine monsters, including the sea-serpent, in their endeavors to attach a precise meaning to the word ; while others have been content to admit that it may signify any kind or all kinds of large aquatic animals. The greater part of the difficulty appears to have arisen from confounding two distinct words, *tannin* and *tan*, both names of animals ; and the confusion has been increased by the circumstance that in two places the words have been interchanged, probably by errors of transcribers. *Tan* occurs in twelve places, and from these we can gather that it inhabits ruined cities, deserts, and places to which ostriches resort, that it suckles

its young, is of predaceous and shy habits, utters a wailing cry, and is not of large size, nor formidable to man. The most probable conjecture as to the animal intended is that of Gesenius, who supposes it to be the jackal. The other word (*tannin*), which is that used in the text, is applied as an emblem of Egypt and its kings, and also of the conquering kings of Babylon. It is spoken of as furious when enraged, and formidable to man, and is said to be an inhabitant of rivers and of the sea, but more especially of the Nile. In short, it is the crocodile of the Nile. We can easily understand the perplexity of those writers who suppose these two words to be identical, and endeavor to combine all the characters above mentioned in one animal or tribe of animals. As a farther illustration of the marked difference in the meanings of the two words, we may compare the 34th and 37th verses of the fifty-first chapter of Jeremiah. In the first of these verses the King of Babylon is represented as a "dragon" (*tannin*), which had swallowed up Israel. In the second it is predicted that Babylon itself shall become heaps, a dwelling-place for "dragons" (*tanim*). There can be no doubt that the animals intended here are quite different. The devouring *tannin* is a huge predaceous river reptile, a fit emblem of the Babylonian monarch ; the *tan* is the jackal that will soon howl in his ruined palaces. It is interesting to know that philologists trace a connection between *tannin* and the Greek *teino*, Latin *tendo*, and similar words, signifying to stretch or extend, in the Sanscrit, Gothic, and other languages, leading to the inference that the Hebrew word primarily denotes a lengthened or extended creature, which corresponds well with its application to the crocodile. Taking all the above facts in connection, we are quite safe in concluding that the creatures referred to by the word under consideration are literally large reptilian animals; and, from the special mention made of them, we may infer that, in their day, they were the lords of creation.*

3. In verse 21 the remainder of the *sheretzim*, besides the larger reptiles, are included in the general expression, "Living creature that moveth." The term "living creature" is, literally, "creature having the breath of life;" the power of respiration being apparently in Hebrew the distinctive character of the animal. The word moveth (*ramash*), in its more general sense, expresses the power of voluntary motion, as exhibited in animals in general. In a few places, however, it has a more precise meaning, as in 1 Kings iv., 33, where the vertebrated animals are included in the four classes of "beasts,' fowl, *creeping things* (or reptiles, *remes*), and fishes." In the present connection it probably has its most general sense; unless, indeed, the apparent repetition in this verse relates to the amphibious or semi-terrestrial creatures associated with the great reptiles; and, in that case, the humbler reptilian animals alone may be meant.

4. We may again note that the introduction of animal life is marked by the use of the word "create," for the first time since the general creation of the heavens and the earth. We may also note that the animal, as well as the plant, was created "after its kind," or "species by species." The animals are grouped under three great classes—the Remes, the Tanninim, and the Birds; but, lest any misconception should arise as to the relations of species to these groups, we are expressly informed that the species is here the true unit of the creative work. It is worth while, therefore, to note that this most an-

---

* It would be unfair to suppress the farther probability that the writer intends specially to indicate that the sacred crocodile of the Nile was itself a creature of Jehovah, and among the humbler of those creatures.

cient authority on this much controverted topic connects species on the one hand with the creative fiat, and on the other with the power of continuous reproduction.

5. In addition to the great mass of *sheretzim*, so accurately characterized by Milton as

" ——Reptile with spawn abundant,"

the creation of the fifth day included a higher tribe of oviparous animals—the birds, the fowl or winged creature of the text. Birds alone, we think, must be meant here, as we have already seen that insects are included under the general term *sheretzim*.

6. It is farther to be observed that *the waters* give origin to the first animals—an interesting point when we consider the contrast here with the creation of plants and of the higher animals, both of which proceed from the earth.

. 7. It can not fail to be observed that we have in these verses two different arrangements of the animals created, neither corresponding exactly with what modern science teaches us to regard as the true grouping of the animal kingdom, according to its affinities. The order in the first enumeration should, from the analogy of the chapter, indicate that of successive creation. The order of the second list may, perhaps, be that of the relative importance of the animals, as it appeared to the writer. Or there may have been a twofold division of the period — the earlier commencing with the creation of the humbler invertebrates, the later characterized by the great reptiles—which is the actual state of the case as disclosed by geology.

8. The Creator recognizes the introduction of sentient existence and volition by *blessing* this new work of his hands, and inviting the swarms of the newly peopled world to enjoy

that happiness for which they were fitted, and to increase and fill the earth, inaugurating thus a new power destined to still higher developments.

When we inquire what information geology affords respecting the period under consideration, the answer may be full and explicit. Geological discovery has carried us back to an epoch corresponding with the beginning of this day, and has disclosed a long and varied series of living beings, extending from this early period up to the introduction of the higher races of animals. To enter on the geological details of these changes, and on descriptions of the creatures which succeeded each other on the earth, would swell this volume into a treatise on palæontology, and would be quite unnecessary, as so many excellent popular works on this subject already exist. I shall, therefore, confine myself to a few general statements, and to marking the points in which Scripture and geology coincide in their respective histories of this long period, which appears to include the whole of the Palæozoic and Mesozoic epochs of geology, with their grand and varied succession of rock formations and living beings.

In the Primordial or oldest fossiliferous rocks next in succession to those great Eozoic formations in which protozoa alone have been discovered, we find the remains of crustaceans, mollusks, and radiates—such as shrimps, shell-fish, and starfishes—which appear to have inhabited the bottom of a shallow ocean. Among these were some genera belonging to the higher forms of invertebrate life, but apparently as yet no vertebrated animals. Fishes were then introduced, and have left their remains in the upper Silurian rocks, and very abundantly in the Devonian and Carboniferous, in the latter of which also the first reptiles occur, but are principally members of that lower group to which the frogs and newts and

K

their allies belong. The animal kingdom appears to have
reached no higher than the reptiles in the Palæozoic or
primary period of geology, and its reptiles are comparatively
small and few ; though fishes had attained to a point of perfection which they have not since exceeded.   There was
also, especially in the Carboniferous age, an abundant and
luxuriant vegetation.   The Mesozoic period is, however, emphatically the age of reptiles.   This class then reached its climax, in the number, perfection, and magnitude of its species,
which filled all those stations in the economy of nature now
assigned to the mammalia.   Birds also belong to this era,
though apparently much less numerous and important than at
present.   Only a few species of small mammals, of the lowest
or marsupial type, appear as a presage of the mammalian creation of the succeeding tertiary era.   In these two geological periods, then—the Palæozoic and Mesozoic—we find, first, the
lower *sheretzim* represented by the invertebrata and the fishes,
then the great reptiles and the birds ; and it can not be denied that, if we admit that the Mosaic day under consideration
corresponds with these geological periods, it would be impossible better to characterize their creations in so few words
adapted to popular comprehension.   I may add that all the
species whose remains are found in the Palæozoic and Mesozoic rocks are extinct, and known to us only as fossils ; and
their connection with the present system of nature consists
only in their forming with it a more perfect series than our
present fauna alone could afford, unless, indeed, we should
find reason to believe that any modern animals are their
modified descendants.   They belong to the same system of
types, but are parts of it which have served their purpose and
have been laid aside.   The coincidences above noted between geology and Scripture may be summed up as follows:

1. According to both records, the causes which at present regulate the distribution of light, heat, and moisture, and of land and water, were, during the whole of this period, much the same as at present. The eyes of the trilobite of the old Silurian rocks are fitted for the same conditions with respect to light with those of existing animals of the same class. The coniferous trees of the coal measures show annual rings of growth. Impressions of rain-marks have been found in the shales of the coal measures and Devonian system. Hills and valleys, swamps and lagoons, rivers, bays, seas, coral reefs and shell beds, have all left indubitable evidence of their existence in the geological record. On the other hand, the Bible affirms that all the earth's physical features were perfected on the fourth day, and immediately before the creation of animals. The land and the water have undergone during this long lapse of ages many minor changes. Whole tribes of animals and plants have been swept away and replaced by others, but the general aspect of inorganic nature has remained the same.

2. Both records show the existence of vegetation during this period; though the geologic record, if taken alone, would, from its want of information respecting the third day, lead us to infer that plants are no older than animals, while the Bible does not speak of the nature of the vegetation that may have existed on the fifth day.

3. Both records inform us that reptiles and birds were the higher and leading forms of animals, and that all the lower forms of animals co-existed with them. In both we have especial notice of the gigantic Saurian reptiles of the latter part of the period; and if we have the remains of a few small species of mammals in the Mesozoic rocks, these, like a few similar creatures apparently included under the word *sherets*

in Leviticus, are not sufficiently important to negative the general fact of the reign of reptiles.*

4. It accords with both records that the work of creation in this period was gradually progressive. Species after species was locally introduced, extended itself, and, after having served its purpose, gradually became extinct. And thus each successive rock formation presents new groups of species, each rising in numbers and perfection above the last, and marking a gradual assimilation of the general conditions of our planet to their present state, yet without any convulsions or general catastrophes affecting the whole earth at once.

5. In both records the time between the creation of the first animals and the introduction of the mammalia as a dominant class forms a well-marked period. I would not too positively assert that the close of the fifth day accords precisely with that of the Mesozoic or secondary period. The well-marked line of sepafation, however, in many parts of the world, between this and the earlier tertiary rocks succeeding to it, points to this as extremely probable.

It thus appears that Scripture and geology so far concur respecting the events of this period as to establish, even without any other evidence, a probability that the fifth day corresponds with the geological ages with which I have endeavored to identify it. Geology, however, gives us no

---

* The interesting discovery, by Mr. Beale and others, of several species of mammalia in the Purbeck, and that of Professor Emmons of a mammal in rocks of similar age in the Southern States of America, do not invalidate this statement ; for all these, like the *Microlestes* of the German trias and the *Amphitherium* of the Stonesfeld slate, are small marsupials belonging to the least perfect type of mammals. The discovery of so many species of these humbler creatures, goes far to increase the improbability of the existence of the higher mammals.

means of measuring precisely the length of this day ; but it gives us the impression that it occupied an enormous length of time, compared with which the whole human period is quite insignificant ; and rivalling those mythical " days of the Creator " which we have noticed as forming a part of the Hindoo mythology.

Why was the earth thus occupied for countless ages by an animal population whose highest members were reptiles and birds ? The fact can not be doubted, since geology and Scripture, the research of man and the Word of God, concur in affirming it. We know that the lowest of these creatures was, in its own place, no less worthy of the Creator than those which we regard as the highest in the scale of organization, and that the animals of the ancient, equally with those of the modern world, abounded in proofs of the wisdom, power, and goodness of their Maker. Comparative anatomy has shown that these extinct animals, though often varying much from their modern representatives, are in no respect rude or imperfect ; that they have the same appearance of careful planning and elaborate execution, the same combination of ornament and utility, the same nice adaptation to the conditions of their existence, which we observe in modern creatures. In addition to this, the many new and wonderful contrivances and combinations which they present, and their relations to existing objects, have greatly enlarged our views of the variety and harmony of the whole system of nature. They are, therefore, in these respects, not without their use as manifestations of the Creator, in this our later age.

There is another reason, hinted at by Buckland, Miller, and other writers on this subject, which weighs much with my mind. All animals and plants are constructed on a few leading types or patterns, which are again divided into sub-

ordinate types, just as in architecture we have certain leading styles, and these again may admit of several orders, and
these of farther modifications. Types are farther modified to
suit a great variety of minor adaptations. Now we know that
the earth is, at any one time, inadequate to display all the
modifications of all the types. Hence our existing system
of organic nature, though probably more complete than any
that preceded it, is still only fragmentary. It is like what
architecture would be, if all memorials of all buildings more
than a century old were swept away. But, from the beginning to the end of the creative work, there has been, or will
be, room for the whole plan. Hence fossils are little by little completing our system of nature ; and, if all were known,
would perhaps wholly do so. The great plan must be progressive, and all its parts must be perishable, except its last
culminating-point and archetype, man. Tennyson expresses
this truth in the following lines :

> "The wish that of the living whole
>   No life may fail beyond the grave ;
>   Derives it not from what we have
> The likest God within the soul?
>
> Are God and Nature then at strife, •
>   That Nature lends such evil dreams ?
>   So careful of the type she seems,
> So careless of the single life.
>
> 'So careful of the type ?' but no.
>   From scarped cliff and quarried stone
>   She cries, 'a thousand types are gone ;
> I care for nothing, all shall go.
>
> 'Thou makest thine appeal to me :
>   I bring to life, I bring to death :
>   The spirit does but mean the breath :
> I know no more.' And he, shall he,

Man, her last work, who seem'd so fair,
    Such splendid purpose in his eyes,
    Who roll'd the psalm to wintry skies,
Who built him fanes of fruitless prayer,

Who trusted God was love indeed,
    And love Creation's final law—
    Tho' Nature, red in tooth and claw,
With ravine, shriek'd against his creed—

Who loved, who suffer'd countless ills,
    Who battled for the True, the Just,
    Be blown about the desert dust,
Or seal'd within the iron hills?

No more? A monster, then, a dream,
    A discord. Dragons of the prime,
    That tare each other in their slime,
Were mellow music match'd with him.

O life as futile, then, as frail!
    O for thy voice to soothe and bless!
    What hope of answer, or redress?
Behind the veil, behind the veil."

The farther explanation given by evolutionists that those ancient forms of life may be the actual ancestors of the present animals, and that through all the ages the Creator was gradually perfecting his work by a series of descents with modification, was probably not before the mind of our ancient Hebrew authority, nor need we attach much value to it till some proof of the process has been obtained from Nature. A farther reason, however, which was intelligible to the author of Genesis, and which is fondly dwelt on in succeeding books of the Bible, depends on the idea that the Creator himself is not indifferent to the marvellous structures, instincts, and powers which he has bestowed upon the lower races of animals.

Witness the answer of the Almighty to Job, when he spake
out of the whirlwind to vindicate his own plans in creation
and providence; and brought before the patriarch a long train
of animals, explaining and dwelling on the structure and powers of each, in contrast with the puny efforts and rude artificial
contrivances of man. Witness also the preservation, in the
rocks, of the fossil remains of extinct creatures, as if he who
made them was unwilling that the evidence of their existence
should perish, and purposely treasured them through all the
revolutions of the earth, that through them men might mag-
nify his name. The Psalmist would almost appear to have
had all these thoughts before his mind when he poured out
his wonder in the 104th Psalm:

> "O Lord, how manifold are thy works!
> In wisdom hast thou made them all.
> The earth is full of thy riches;
> So is this wide and great sea,
> Wherein are moving things innumerable,
> Creatures both small and great.
> There go the ships [or "floating animals"];
> There is leviathan, which thou hast formed to sport therein:
> That thou givest them they gather.
> Thou openest thy hand, they are filled with good;
> Thou hidest thy face, they are troubled;
> Thou takest away their breath, they return to their dust.
> Thou sendest forth thy spirit, they are created,
> And thou renewest the face of the earth."

There are, however, good reasons to believe that, in the
plans of divine wisdom, the long periods in which the earth
was occupied by the inferior races were necessary to its subsequent adaptation to the residence of man. In these periods
our present continents gradually grew up in all their variety
and beauty. The materials of old rocks were comminuted and

mixed to form fertile soils,* and stores of mineral products were accumulated to enable man to earn his subsistence and the blessings of civilization by the sweat of his brow. If it pleased the Almighty during these preparatory stages to re- plenish the land and sea with living things full of life and beauty and happiness, who shall venture to criticise his pro- cedure, or to say to Him, " What doest thou?"

It would be decidedly wrong, in the present state of that which is popularly called science, to omit to inquire here what relation to the work of the fifth creative day those theories of development and evolution which have obtained so great cur- rency may bear. The long time employed in the introduc- tion of the lower animals, the use of the terms " make " and " form," instead of " create," and the expression " let the wa- ters bring forth," may well be understood as countenancing some form of mediate creation, or of " creation by law," or " theistic evolution," as it has been termed ; but they give no countenance to the idea either of the spontaneous evolu- tion of living beings under the influence of merely physical causes and without creative intervention, or of the transmuta- tion of one kind of animal into another. Still, with reference to this last idea, it is plain that revelation gives us no defini- tion of species as distinguished from varieties or races, so that there is nothing to prevent the supposition that, within certain limits indicated by the expression " after its kind," animals or plants may have been so constituted as to vary greatly in the progress of geological time.

If we ask whether any thing is known to science which can

---

* It is very interesting, in connection with this, to note that nearly all the earliest and greatest seats of population and civilization have been placed on the more modern geological deposits, or on those in which stores of fuel have been accumulated by the growth of extinct plants.

K 2

give even a decided probability to the notion that living beings are parts of an undirected evolution proceeding under
merely dead insentient forces, and without intention, the answer must be emphatically no.

I have elsewhere fully discussed these questions, and may
here make some general statements as to certain scientific
facts which at present bar the way against the hypothesis of
evolution as applied to life, and especially against that form
of it to which Darwin and his disciples have given so great
prominence.

1. The albuminous or protoplasmic material, which seems
to be necessary to the existence of every living being, is known
to us as a product only of the action of previously living protoplasm.   Though it is often stated that the production of
albumen from its elements is a process not differing from the
formation of water or any other inorganic material from its
elements, this statement is false in fact, since, though many
so-called organic substances have been produced by chemical
processes, no particle of either living or non-living organizable matter of the nature of protoplasm has ever been so produced.   The origin, therefore, of this albuminous matter is
as much a mystery to us at present as that of any of the
chemical elements.

2. Though some animals and plants are very simple in
their visible structure, they all present vital properties not to
be found in dead albuminous matter, and no mode is known
whereby the properties of life can be communicated to dead
matter.   All the experiments hitherto made, and very eminently those recently performed by Pasteur, Tyndall, and Dal-
linger, lead to the conclusion that even the simplest living
beings can be produced only from germs originating in previously living organisms of similar structure.   The simplest

living organisms are thus to science ultimate facts, for which it can not account except conjecturally.

3. No case is certainly known in human experience where any species of animal or plant has been so changed as to assume all the characters of a new species. Species are thus practically to science unchangeable units, the origin of which we have as yet no means of tracing.

4. Though the general history of animal life in time bears a certain resemblance to the development of the individual animal from the embryo, there is no reason whatever to believe that this is more than a mere relation of analogy, arising from the fact that in both cases the law of procedure is to pass from the simpler forms to the more complex, and from the more generalized to the more specialized. The external conditions and details of the two kinds of series are altogether different, and become more so the more they are investigated. This shows that the causes can not have been similar.

5. In tracing back animals and groups of animals in geological time, we find that they always end without any link of connection with previous beings, and in circumstances which render any such connections improbable. In the work of our next creative day, the series of animals preceding the modern horse has been cited as a good instance of probable evolution ; but not only are the members of the series so widely separated in space and time that no connection can be traced, but the earliest of them, the *Orohippus*, would require, on the theory, to have been preceded by a previous series extending so far back that it is impossible, under any supposition of the imperfection of our present knowledge, to consider such extension probable. The same difficulty applies to every case of tracing back any specific form either of animal or plant. This general result proves, as I have elsewhere attempted to

show,* that the introduction of the various animal types must
have been abrupt, and under some influence quite different
from that of evolution.

⌐ These are what I would term the five fatal objections to
evolution as at present held, as a means of accounting for the
introduction and succession of animals. To what extent they
may be weakened or strengthened by the future progress of
science it is impossible to say, but so long as they exist it is
mere folly and presumption to affirm that modern science supports the doctrine of evolution. There can be no doubt, however, that the Bible leaves us perfectly free to inquire as to
the plan and method of the Creator, and that, whatever discoveries we may make, we shall find that his plans are orderly, methodical, and continuous, and not of the nature of an
arbitrary patchwork.

Though science as yet gives us no certain laws for the introduction of new specific types, it indicates certain possible
modes of the origination of varieties, races, and sub-species
of previously existing types. One of these is that struggle for
existence against adverse external conditions, which, however,
has been harped upon too exclusively by the Darwinian school,
and which will give chiefly depauperated and degraded forms.
Another is that expansion under exceptionally favorable conditions which arises where species are admitted to wider new
areas of geographical range and more abundant and varied
means of sustenance. Land animals and plants must have
experienced this in times of continental elevation ; marine
animals and plants in times of continental depression. Another is the tendency to what has been called reproductive
retardation and acceleration which species undergo under

* See Appendix.

conditions exceptionally unfavorable or favorable, and which in some modern aquatic animals produces differences so great that members of the same species have sometimes been placed in different genera. Lastly, it is conceivable that species may have been so constructed that after a certain number of generations they may spontaneously undergo either abrupt or gradual changes, similar to those which the individual undergoes at certain stages of growth. This last furnishes the only true analogy possible between embryology and geological succession.

While, however, science is silent as to the production of new specific types, and only gives us indications as to the origin of varieties and races, it is curious that the Bible suggests three methods in which new organisms may be, and according to it have been introduced by the Creator. The first is that of immediate and direct creation, as when God created the great Tanninim. The second is that of mediate creation, through the materials previously existing, as when he said, "Let the land bring forth plants," or "Let the waters bring forth animals." The third is that of production from a previous organism by power other than that of ordinary reproduction, as in the origination of Eve from Adam, and the miraculous conception of Jesus. These are the only points in which its teachings approach the limits of speculations as to evolution, and they certainly leave scope enough for the legitimate inquiries of science.*

---

* See Appendix for farther discussion of this subject.

CHAPTER XI.

THE HIGHER ANIMALS AND MAN.

" And God said, Let the land bring forth animals after their kinds; the herbivora, the reptiles, and the carnivora, after their kinds; and it was so. And God made carnivorous mammals after their kinds, and herbivorous mammals after their kinds, and every reptile of the land after its kind; and God saw that it was good.

" And God said, Let us make man in our own image, after our likeness; and let them rule over the fish of the sea and the birds of the air, and over the herbivora and over all the land. So God created man in his own image, in the image of God created he him; male and female created he them. And God blessed them; and God said, Be fruitful and multiply, and replenish the earth, and subdue it; and have dominion over the fish of the sea and over the fowl of the air, and over every living thing that moveth upon the earth.

" And God said, Behold, I have given you every herb bearing seed which is upon the face of all the earth, and every tree in which is the fruit of a tree yielding seed; to you it shall be for food, and to every beast of the earth and to every fowl of the air, and to every thing that creepeth upon the earth wherein there is life, I have given every green herb for meat; and it was so. And God saw every thing that he had made, and, behold, it was very good. And evening and morning were the sixth day."—Genesis i., 24-31.

THE creation of animals, unlike that of plants, occupies two days. Here our attention is restricted to the inhabitants of the *land*, and chiefly to their higher forms. Several new names are introduced to our notice, which I have endeavored to translate as literally as possible by introducing zoological terms where those in common use were deficient.

1. The first tribe of animals noticed here is named
Bhemah, "cattle" in our version; and in the Septuagint
"quadrupeds" in one of the verses, and "cattle" in the
other. Both of these senses are of common occurrence in
the Scriptures, cattle or domesticated animals being usually
designated by this word; while in other passages, as in
1 Kings iv., 33, where Solomon is said to have written a
treatise on "*beasts*, fowls, creeping things, and fishes," it
appears to include all the mammalia. Notwithstanding this
wide range of meaning, however, there are passages, and
these of the greatest authority in reference to our present
subject, in which it strictly means the herbivorous mammals,
and which show that when it was necessary to distinguish
these from the predaceous or carnivorous tribes this term
was specially employed. In Leviticus xi., 22–27, we have a
specification of all the Bhemoth that might and might not be
used for food. It includes all the true ruminants, with the
coney, the hare, and the hog, animals of the rodent and pachy-
dermatous orders. The carnivorous quadrupeds are designated by a different generic term. In this chapter of Leviticus, therefore, which contains the only approach to a system
in natural history to be found in the Bible, *bhemah* is strictly
a synonym of *herbivora*, including especially ungulates and
rodents. That this is its proper meaning here is confirmed
by the considerations that in this place it can denote but a
part of the land quadrupeds, and that the idea of cattle or
domesticated animals would be an anachronism. At the
same time there need be no objection to the view that the
especial capacity of ruminants and other herbivora for domestication is connected with the use of the word in this place.

2. The word *remes*, "creeping things" in our version, as we
have already shown, is a very general term, referring to the

power of motion possessed by animals, especially on the surface of the ground.   It here in all probability refers to the additional types of terrestrial reptiles, and other creatures lower than the mammals, introduced in this period.

3. The compound term (*hay'th-eretz*) which I have ventured to render "carnivora," is literally animal of the land; but though thus general in its meaning, it is here evidently intended to denote a particular tribe of animals inhabiting the land, and not included in the scope of the two words already noticed.   In other parts of Scripture this term is used in the sense of a "wild beast."   In a few places, like the other terms already noticed, it is used of all kinds of animals, but that above stated is its general meaning, and perfectly accords with the requirements of the passage.

The creation of the sixth day therefore includes—1st, the herbivorous mammalia; 2d, a variety of terrestrial reptilia, and other lower forms not included in the work of the previous day; 3d, the carnivorous mammalia.   It will be observed that the order in the two verses is different.   In verse 24th it is herbivora, "creeping things," and carnivora.   In verse 25th it is carnivora, herbivora, and "creeping things." One of these may, as in the account of the fifth day, indicate the order of *time* in the creation, and the other the order of *rank* in the animals made, or there may have been two divisions of the work, in the earlier of which herbivorous animals took the lead, and in the later those that are carnivorous. In either case we may infer that the herbivora predominated in the earlier creations of the period.

It is almost unnecessary to say this period corresponds with the Tertiary or Cainozoic era of geologists.   The coincidences are very marked and striking.   As already stated, though in the later secondary period there were great facilities for the preservation of mammals in the strata then being deposited, only a few small species of the humblest order have been found; and the occurrence of the higher orders of this class is to some extent precluded by the fact that the place in nature now occupied by the mammals was then provided for by the vast development of the reptile tribes. At the very beginning of the tertiary period all this was changed; most of the gigantic reptiles had disappeared, and terrestrial mammals of large size and high organization had taken their place. Perhaps no geological change is more striking and remarkable than the sudden disappearance of the reptilian fauna at the close of the mesozoic, and the equally abrupt appearance of numerous species of large mammals, and this not in one region only, but over both the great continents, and not only where a sudden break occurs in the series of formations, but also where, as in Western America, they pass gradually into each other. During the whole tertiary period this predominance of the mammalia continued; and as the mesozoic was the period of giant reptiles, so the tertiary was that of great mammals. It is a singular and perhaps not accidental coincidence that so many of the early tertiary mammals known to us are large herbivora, such as would be included in the Hebrew word *bhemah;* and that in the book of Job the hippopotamus is called *behemoth*, the plural form being apparently used to denote that this animal is the chief of the creatures known under the general term *bhemah*, while geology informs us that the prevailing order of mammals in the older tertiary period was that of the ungulates, and that many of the extinct creatures of this group are very closely allied to the hippopotamus. Behemoth thus figures in the book of Job, not only as at the time a marked illustration of creative power, but to our farther knowledge also as a singular remnant of an extinct gigantic race. It is at least curious that while in the fifth day great reptiles like those of the secondary rocks form the burden of the work, in the sixth we have a term which so directly reminds us of those gigantic pachyderms which figure so largely in the tertiary period. Large carnivora also occur in the tertiary formations, and there are some forms of reptile life, as, for example, the serpents, which first appear in the tertiary.

I may refer to any popular text-book of geology in evidence of the exact conformity of this to the progress of mammalian life, as we now know it in detail from the study of the successive tertiary deposits. The following short summary from Dana, though written several years ago, still expresses the main features of the case :

"The quadrupeds did not all come forth together. Large and powerful herbivorous species first take possession of the earth, with only a few small carnivora. These pass away. Other herbivora with a larger proportion of carnivora next appear. These also are exterminated; and so with others. Then the carnivora appear in vast numbers and power, and the herbivora also abound. Moreover these races attain a magnitude and number far surpassing all that now exist, as much so indeed, on all the continents, North and South America, Europe, Asia, Africa, and Australia, as the old mastodon, twenty feet long and nine feet high, exceeds the modern buffalo. Such, according to geology, was the age of mammals, when the brute species existed in their greatest magnificence, and brutal ferocity had free play; when the dens of bears and hyenas, prowling tigers and lions far larger than any now existing, covered Britain and Europe. Mammoths and mastodons wandered over the plains of North America, huge sloth-like Megatheria passed their sluggish lives on the pampas of South America, and elephantine marsupials strolled about Australia.

" As the mammalian age draws to a close, the ancient carnivora and herbivora of that era all pass away, excepting, it is believed, a few that are useful to man. New creations of smaller size peopled the groves; the vegetation received accessions to its foliage, fruit-trees and flowers, and the seas brighter forms of water life. This we know from comparisons with the fossils of the preceding mammalian age. There was at this time no chaotic upturning, but only the opening of creation to its fullest expansion ; and so in Genesis no new day is begun, it is still the *sixth day.*"

The creation of man is prefaced by expressions implying deliberation and care. It is not said, " Let the earth bring forth " man, but let us form or fashion man. This marks the relative importance of the human species, and the heavenly origin of its nobler immaterial part. Man is also said to have been "created," implying that in his constitution there was something new and not included in previous parts of the work, even in its material. Man was created, as the Hebrew literally reads, the shadow and similitude of God—the greatest of the visible manifestations of Deity in the lower world—the reflected image of his Maker, and, under the Supreme Lawgiver, the delegated ruler of the earth. Now for the first time was the earth tenanted by a being capable of comprehending the purposes and plans of Jehovah, of regarding his works with intelligent admiration, and of shadowing forth the excellences of his moral nature. For countless ages the earth had been inhabited by creatures wonderful in their structures and instincts, and mutely testifying, as their buried remains still do, to the Creator's glory ; but limited within a narrow range of animal propensities, and having no power of raising a thought or aspiration toward the Being who made them. Now, however, man enters on the scene, and the sons of God, who had shouted for joy when the first land emerged from the bosom of the deep, saw the wondrous spectacle of a spiritual nature analogous to their own, united to a corporeal frame constructed on the same general type with the higher of those irrational creatures whose presence on earth they had so long witnessed.

Man was to rule over the fish of the sea, the birds of the air, and the *bhemah* or herbivorous animals. The carnivorous creatures are not mentioned, and possibly were not included in man's dominion. We shall find an explanation of this farther on. The nature of man's dominion we are left to infer. In his state of innocence it must have been a mild and gentle sway, interfering in no respect with the free exercise of the powers of enjoyment bestowed on animals by the Creator, a rule akin to that which a merciful man exercises over a domesticated animal, and which some animals are capable of repaying with a warm and devoted affection. Now, however, man's rule has become a tyranny. "The whole creation groans" because of it. He desolates the face of nature wherever he appears, unsettling the nice balance of natural agencies, and introducing remediless confusion and suffering among the lower creatures, even when in the might of his boasted civilization he professes to renovate and improve the face of nature. He retains enough of the image of his Maker to enable him to a great extent to assert his dominion, and to aspire after a restoration of his original paradise, but he has lost so much that the power which he retains is necessarily abused to selfish ends.

Man, like the other creatures, was destined to be fruitful and multiply and replenish the earth. We are also informed

in chapter second that he was placed in a "garden," a chosen
spot in the alluvial plains of Western Asia, belonging to the
later geological formations, and thus prepared by the whole
series of prior geological changes, replenished with all things
useful to him, and containing nothing hurtful, at least in so
far as the animal creation was concerned.   These facts,
taken in connection, lead to grave questions.   How is the
happy and innocent state of man consistent with the contemporaneous existence of carnivorous and predaceous animals, which, as both Scripture and geology state, were created
in abundance in the sixth day?   How, when confined to a
limited region, could he increase and multiply and replenish
the earth?   These questions, which have caused no little
perplexity, are easily solved when brought into the light of
our modern knowledge of nature.   1. Every large region of
the earth is inhabited by a group of animals differing in the
proportions of identical species, and in the presence of distinct
species, from the groups inhabiting other districts.   There is
also sufficient reason to conclude that all animals and plants
have spread from certain local centres of creation, in which
certain groups of species have been produced and allowed
to extend themselves, until they met and became intermingled with species extending from other centres.   Now
the district of Asia, in the vicinity of the Euphrates and
Tigris, to which the Scripture assigns the origin of the human race, is the centre to which we can with the greatest
probability trace several of the species of animals and plants
most useful to man, and it lies near the confines of warmer
and colder regions of distribution in the Old World, and also
near the boundary of the Asiatic and European regions.   At
the period under consideration it may have been peopled
with a group of animals specially suited to association with

the progenitors of mankind. 2. To remove all zoological difficulties from the position of primeval man in his state of innocence, we have but to suppose, in accordance with all the probabilities of the case, that man was created along with a group of creatures adapted to contribute to his happiness, and having no tendency to injure or annoy; and that it is the formation of these creatures—the group of his own centre of creation—that is especially noticed in Genesis ii., 19, *et seq.*, where God is represented as forming them out of the ground and exhibiting them to Adam ; a passage otherwise superfluous, and indeed tending to confuse the meaning of the document. 3. The difficulty attending the early extension of the human race is at once obviated by the geological doctrine of the extinction of species. We know that in past geological periods large and important groups of species have become extinct, and have been replaced by new groups extending from new centres ; and we know that this process has removed, in early geological periods, many creatures that would have been highly injurious to human interests had they remained. Now the group of species created with man being the latest introduced, we may infer, on geological grounds, that it would have extended itself within the spheres of older zoological and botanical districts, and would have replaced their species, which, in the ordinary operation of natural laws, may have been verging toward extinction. Thus not only man, but the Eden in which he dwelt, with all its animals and plants, would have gradually encroached on the surrounding wilderness, until man's happy and peaceful reign had replaced that of the ferocious beasts that preceded him in dominion, and had extended at least over all the temperate region of the earth. 4. The cursing of the ground for man's sake, on his fall from innocence, would thus consist in the permission given to the predaceous animals and the thorns and the briers of other centres of creation to invade his Eden ; or, in his own expulsion, to contend with the animals and plants which were intended to have given way and become extinct before him. Thus the fall of man would produce an arrestment in the progress of the earth in that last great revolution which would have converted it into an Eden ; and the anomalies of its present state consist, according to Scripture, in a mixture of the conditions of the tertiary with those of the human period. 5. Though there is good ground for believing that man was to have been exempted from the general law of mortality, we can not infer that any such exemption would have been enjoyed by his companion animals ; we only know that he himself would have been free from all annoyance and injury and decay from external causes. We may also conclude that, while Eden was sufficient for his habitation, the remainder of the earth would continue, just as in the earlier tertiary periods, under the dominion of the predaceous mammals, reptiles, and birds. 6. The above views enable us on the one hand to avoid the difficulties that attend the admission of predaceous animals into Eden, and on the other the still more formidable difficulties that attend the attempt to exclude them altogether from the Adamic world. They also illustrate the geological fact that many animals, contemporaneous with man, extend far back into the Tertiary period. These are creatures not belonging to the Edenic centre of creation, but introduced in an earlier part of the sixth day, and now permitted to exist along with man in his fallen state. I have stated these supposed conditions of the Adamic creation briefly, and with as little illustration as possible, that they may connectedly strike the mind of the reader.

Each of these statements is in harmony with the Scriptural narrative on the one hand, and with geology on the other; and, taken together, they afford an intelligible history of the introduction of man.   If a geologist were to state, *à priori*, the conditions proper to the creation of any important species, he could only say—the preparation or selection of some region of the earth for it, and its production along with a group of plants and animals suited to it.   These are precisely the conditions implied in the Scriptural account of the creation of Adam.*   The difficulties of the subject have arisen from supposing, contrary to the narrative itself, that the conditions necessary for Eden must in the first instance have extended over the whole earth, and that the creatures with which man is in his present dispersion brought into contact must necessarily have been his companions there. One would think that many persons derive their idea of the first man in Eden from nursery picture-books ; for the Bible gives no countenance to the idea that all the animals in the world were in Eden.   On the contrary, it asserts that a selection was made both in the case of animals and plants, and that this Edenic assemblage of creatures constituted man's associates in his state of primeval innocence.

The food of animals is specified at the close of the work of this day.   The grant to man is every herb bearing seed, and every fruit-tree.   That to the lower animals is more extensive—every green herb.   This can not mean that every animal in the earth was herbivorous.   It may refer to the group of animals associated with man in Eden, and this is most likely the intention of the writer ; but if it includes the animals of the whole earth, we may be certain, from the ex-

---

* See Lyell, Principles of Geology, "Introduction of Species."

press mention of carnivorous creatures in the work of the fifth and sixth days, that it indicates merely the general fact that the support of the whole animal kingdom is based on vegetation.

A most important circumstance in connection with the work of the sixth day is that it witnessed the creation both of man and the mammalia. A fictitious writer would probably have exalted man by assigning to him a separate day, and by placing the whole animal kingdom together in respect to time. He would be all the more likely to do this, if unacquainted, as most ignorant persons as well as literary men are, with the importance and teeming multitudes of the lower tribes of animals, and with the typical identity of the human frame with that of the higher animals. Moses has not done so, we are at liberty to suppose, because the vision of creation had it otherwise; and modern geology has amply vindicated him in this by its disclosure of the intimate connection of the human with the tertiary period; and has shown in this as in other instances that truth and not "accommodation" was the object of the sacred writer. While, as already stated, many existing species extend far back into the tertiary period, showing that the earth has been visited by no universal catastrophe since the first creation of mammals; on the other hand, we can not with certainty trace any existing species back beyond the commencement of the tertiary era. Geology and revelation, therefore, coincidé in referring the creation of man to the close of the period in which mammals were introduced and became predominant, and in establishing a marked separation between that period and the preceding one in which the lower animals held undisputed sway. This coincidence, while it strengthens the probability that the creative days were long periods, opposes an almost insur-

mountable obstacle to every other hypothesis of reconciliation with geological science.

At the close of this day the Creator again reviews his work,
and pronounces it good.   Step by step the world had been
evolved from a primeval chaos, through many successive physical changes and long series of organized beings.   It had now
reached its acme of perfection, and had received its most
illustrious tenant, possessing an organism excelling all others
in majesty and beauty, and an immaterial soul the shadow
of the glorious Creator himself.   Well might the angels sing,
when the long-protracted work was thus grandly completed :

> "Thrice happy man,
> And sons of men, whom God hath thus advanced,
> Created in his image, there to dwell
> And worship him, and in reward to rule
> Over his works in earth, or sea, or air,
> And multiply a race of worshippers
> Holy and just ; thrice happy, if they know
> Their happiness and persevere upright."

The Hebrew idea of the golden age of Eden is pure and
exalted.   It consists in the enjoyment of the favor of God,
and of all that is beautiful and excellent in his works.   God
and nature are the whole.   Nor is it merely a rude, unintel-
ligent, sensuous enjoyment.   Man primeval is not a lazy
savage gathering acorns.   He is made in the image of the
Creator ; he is to keep and dress his garden, and it is furnished with every plant good for food and pleasant to the
sight.   In the midst of our material civilization we need to
disabuse ourselves of some prejudices before we can realize
the fact that man, without the arts of life or any need of them,
is not necessarily a barbarian or a savage.   Yet even Adam
must have been an agriculturist with strong and willing

hands, and must have had some need of agricultural implements such as those with which the least civilized of his descendants have been wont to till the soil. Still, without art or with very little of it, he could enjoy all that is beautiful and grand in nature, and could rise from the observation of nature to communion with God. We need the more to realize this, inasmuch as there seems so strong a tendency to confound material civilization with higher culture, and to hold that man primeval must have been low and debased simply because he may have had no temples and no machinery. We must remember that he had nature, which is higher than fine art, and that when in harmony with his surroundings he may have had no need either of exhausting labor or of mechanical contrivances. Farther, in the contemplation of nature and in seeking after God, he had higher teachers than our boasted civilization can claim.

Alas for fallen man, with his poor civilization gathered little by little from the dust of earth, and his paltry art that halts immeasurably behind nature. How little is he able even to appreciate the high estate of his great ancestor. The world of fallen men has worshipped art too much, reverenced and studied God and nature too little. The savage displays the lowest taste when he admires the rude figures which he paints on his face or his garments more than the glorious painting that adorns nature; yet even he acknowledges the pre-eminent excellence of nature by imitating her forms and colors, and by adapting her painted plumes and flowers to his own use. There is a wide interval, including many gradations, between this low position and that of the cultivated amateur or artist. The art of the latter makes a nearer approach to the truly beautiful, inasmuch as it more accurately represents the geometric and organic forms and the coloring of nature; and inasmuch as it devises ideal combinations not found in the actual world; which ideal combinations, however, are beautiful or monstrous just as they realize or violate the harmonies of nature. It is only the highest culture that brings man back to his primitive refinement. ⌒Art takes her true place when she sits at the feet of nature, and brings her students to drink in its beauties, that they may endeavor, however imperfectly, to reproduce them. On the other hand, the student of nature must not content himself with "writing Latin names on white paper," wherewith to label nature's productions, but must rise to the contemplation of the order and beauty of the Cosmos as a revelation of Divinity. Both will thus rise to that highest taste which will enable them to appreciate not only the elegance of individual forms, but their structure, their harmonies, their grouping and their relations, their special adaptation, and their places as parts of a great system. Thus art will attain that highest point in which it displays original genius, without violating natural truth and unity, and nature will be regarded as the highest art.

Much is said and done in our time with reference to the cultivation of popular taste for fine art as a means of civilization; and this, so far as it goes, is well; but the only sure path to the highest taste-education is the cultivation of the study of nature. This is also an easier branch of education, provided the instructors have sufficient knowledge. Good works of art are rare and costly; but good works of nature are everywhere around us, waiting to be examined. Such education, popularly diffused, would react on the efforts of art. It would enable a widely extended public to appreciate real excellence, and would cause works of art to be valued just in proportion to the extent to which they realize or deviate from natural truth and unity. I do not profess to speak authoritatively on such subjects, but I confess that the strong impression on my mind is that neither the revered antique models, nor the practice and principles of the generality of modern art reformers, would endure such criticism; and that if we could combine popular enthusiasm for art with scientific appreciation of nature, a new and better art might arise from the union.

I may appear to dwell too long upon this topic; but my excuse must be that it leads to a true estimate both of natural history and of the sacred Scriptures. The study of nature guides to those large views of the unity and order of creation which alone are worthy of a being of the rank of man, and which lead him to adequate conceptions of the Creator; but the truly wise recognize three grades of beauty. First, that of art, which, in its higher efforts, can raise ordinary minds far above themselves. Secondly, that of nature, which, in its most common objects, must transcend the former, since its artist is that God of whose infinite mind the genius of the artist is only a faint reflection. Thirdly, that pre-eminent beauty of moral goodness revealed only in the spiritual nature of the Supreme. The first is one of the natural resources of fallen man in his search for happiness. The second was man's joy in his primeval innocence. The third is the inheritance of man redeemed. It is folly to place these on the same level. It is greater folly to worship either or both of the first without regard to the last. It is true wisdom to aspire to the last, and to regard nature as the handmaid of piety, art as but the handmaid of nature.

Nature to the unobservant is merely a mass of things more or less beautiful or interesting, but without any definite order or significance. An observer soon arrives at the con-

clusion that it is a series of circling changes, ever returning
to the same points, ever renewing their courses, under the
action of invariable laws. But if he rests here, he falls infinitely short of the idea of the Cosmos, and stands on the
brink of the profound error of eternal succession. A little
further progress conducts him to the inviting field of special
adaptation and mutual relation of things. He finds that
nothing is without its use ; that every structure is most nice-
ly adjusted to special ends ; that the supposed ceaseless circling of nature is merely the continuous action of great powers, by which an infinity of utilities are worked out—the
great fly-wheel which, in its unceasing and at first sight apparently aimless round, is giving motion to thousands of reels
and spindles and shuttles, that are spinning and weaving, in
all its varied patterns, the great web of life.

But the observer, as he looks on this web, is surprised to
find that it has in its whole extent a wondrous pattern. He
rises to the contemplation of type in nature, a great truth to
which science has only lately opened its eyes. He begins
dimly to perceive that the Creator has from the beginning
had a plan before his mind, that this plan embraced various
types or patterns of existence; that on these patterns he has
been working out the whole system of nature, adapting each
to all the variety of uses by an infinity of minor modifications.
That, in short, whether he study the eye of a gnat or the
structure of a mountain chain, he sees not only objects of
beauty and utility, but parts of far-reaching plans of infinite
wisdom, by which all objects, however separated in time or
space, are linked together.

How much of positive pleasure does that man lose who
passes through.life absorbed with its wants and its artifici-
alities, and regarding with a "brute, unconscious gaze" the

grand revelation of a higher intelligence in the outer world. It is only in an approximation through our Divine Redeemer to the moral likeness of God that we can be truly happy; but of the subsidiary pleasures which we are here permitted to enjoy, the contemplation of nature is one of the best and purest. It was the pleasure, the show, the spectacle prepared for man in Eden, and how much true philosophy and taste shine in the simple words that in paradise God planted trees "pleasant to the sight," as well as "good for food." Other things being equal, the nearer we can return to this primitive taste, the greater will be our sensuous enjoyment, the better the influence of our pleasures on our moral nature, because they will then depend on the cultivation of tastes at once natural and harmless, and will not lead us to communion with and reverence for merely human genius, but will conduct us into the presence of the infinite perfection of the Creator.

The Bible knows but one species of man. It is not said that men were created after their species, as we read of the groups of animals. Man was made, "male and female;" and in the fuller details afterwards given in the second chapter—where the writer, having finished his general narrative, commences his special history of man—but one primitive pair is introduced to our notice. We scarcely need the detailed tables of affiliation afterward given, or the declaration of the apostle who preached to the supposed autochthones of Athens, that "God has made of one blood all nations," to assure us of the Scriptural unity of man. If, therefore, there were any good reason to believe that man is not of one but several origins, we must admit Moses to have been very imperfectly informed. Nor, on the other hand, does the Bible any more than geology allow us to assign a very high antiquity to the origin of man relatively to that of the earth on which he dwells. The genealogical tables of the Bible may admit of some limits of difference of opinion as to the age of the human world or æon, and also of that of the deluge, from which man took his second point of departure; but they do not allow us to put the origin of man farther back than that of the present or modern condition of our continents and the present races of animals. They therefore limit us to the modern or quaternary period of geology. The question of man's antiquity, so much agitated now, demands, however, a separate and careful consideration; but we must first devote a few pages to the simple statements of the Bible respecting the Sabbath of creation and its relation to human history.

CHAPTER XII.

THE REST OF THE CREATOR.

"And the heavens and the earth were finished, and all the host of them. And on the seventh day God ended his work which he had made, and he rested on the seventh day from all his work which he had made. And God blessed the seventh day and sanctified it, because that in it God rested from all his work which he had created to make."—Genesis ii., 1-3.

THE end of the sixth day closed the work of creation properly so called, as well as that of forming and arranging the things created. The beginning of the seventh introduced a period which, according to the views already stated, was to be occupied by the continued increase and diffusion of man and the creatures under his dominion, and by the gradual disappearance of tribes of creatures unconnected with his well-being.

Science in this well accords with Scripture. No proof exists of the production of a new species since the creation of man; and all geological and archæological evidence points to him and a few of the higher mammals as the newest of the creatures. There is, on the other hand, good evidence that several species have become extinct since his creation. Those who believe in the continuous evolution of animals and men, it is true, can see no actual termination of the process with the introduction of man; but even they see that the appearance of a rational and moral being at least changes the nature and order of the development. Nor can they doubt that man is the last born of nature, and that the whole animal creation is crowned by him as its capital or topmost pinnacle.   The later speculators on this subject have never reached any truth beyond that long ago stated by the lamented Edward Forbes — a most careful observer and accurate reasoner on the more recent changes of the earth's surface. He infers, from the distribution of species from their centres of creation, that man is the latest product of creative power; or, in other words, that none of those species or groups of species which he had been able to trace to their centres, or the spots at which they probably originated, appear to be of later or as late origin as man.   "This consideration," he says, "induces me to believe that the last province in time was completed by the coming of man, and to maintain an hypothesis that man stands unique in space and time, himself equal to the sum of any pre-existing centre of creation or of all—an hypothesis consistent with man's moral and social position in the world."

The seventh day, then, was to have been that in which all the happiness, beauty, and perfection of the others were to have been concentrated.   But an element of instability was present in the being who occupied the summit of the animal scale.   Not regulated by blind and unerring instincts, but a free agent, with a high intellectual and moral nature, and liable to be acted on by temptation from without; under such influence he lost his moral balance in stretching out his hand to grasp the peculiar powers of Deity, and fell beyond the hope of self-redemption—perpetuating, by one of those laws which regulate the transmission of mixed corporeal and spiritual natures, his degradation to every generation of his species. And so God's great work was marred, and all his plans seemed to be foiled, when they had just reached their completion.

Thus far science might carry us unaided; for there is not a true naturalist, however skeptical as to revealed religion, who does not feel in his inmost heart the disjointed state of the present relations of man to nature; the natural wreck that results from his artificial modes of life, the long trains of violations of the symmetry of nature that follow in the wake of his most boasted achievements.† But here natural science stops; and just as we have found that, in tracing back the world's history, the Bible carries us much farther than geology, so science, having led us to suspect the fallen state of man, leaves us henceforth to the teaching of revelation. And how glorious that teaching!— God did not find himself baffled —his resources are infinite—he had foreseen and prepared for all this apparent evil; and out of the moral wreck he proceeds to work out the grand process of *redemption*, which is the especial object of the seventh day, and which will result in the production of a new heaven and a new earth wherein dwelleth righteousness. In the seventh, as in the former days, the evening precedes the morning. For four thousand years the world groped in its darkness—a darkness tenanted by moral monsters as powerful and destructive as the old pre-Adamite reptiles. The Sun of Righteousness at length arose, and the darkness began to pass away; but eighteen centuries have elapsed, and we still see but the gray dawn of morning, which we yet firmly believe will brighten into a glorious day that shall know no succeeding night.*

The seventh day is the modern or human era in geology; and, though it can not yet boast of any physical changes so

---

* For the exposition of the details of the fall, I beg to refer the reader to McDonald's " Creation and the Fall," to Kitto's " Antediluvians and Patriarchs," and to Kurtz's " History of the Old Covenant."

great as those of past periods, it is still of much interest, as affording the facts on which we must depend for explanations of past changes ; and as immediately connected in time with those later tertiary periods which afford so many curious problems to the geological student. The actual connection of the human with preceding periods is still involved in some obscurity ; and, as we shall see, there has recently been a strong tendency to throw back the origin of man into pre-historic ages of enormous length, on grounds which are, how-ever, much less certain than is commonly imagined. This question we have to examine ; but before entering upon it may shortly sketch the actual import of the statements of the Hebrew Scriptures respecting what may be called the prehis-toric duration of the human species. This is the more neces-sary, as the most crude notions seem very widely to prevail on the subject. I shall, therefore, in this place notice some general facts deducible from the Bible, and which may be useful in appreciating the true relation of the human era to those which preceded it. It will be understood that I shall endeavor merely to present a picture of what the Bible actually teaches, and which any one can verify by reading the book of Genesis.

1. The local centre of creation of the human species, and probably of a group of creatures coeval with it, was Eden ; a country of which the Scriptures give a somewhat minute geo-graphical description. It was evidently a district of Western Asia ; and, from its possession of several important rivers, rather a region or large territory than a limited spot, such as many, who have discussed the question of the site of Eden, seem to suppose. In this view it is a matter of no moment to fix its site more nearly than the indication of the Bible that it included the sources and probably large portions of the val-

leys of the Tigris, the Euphrates, and perhaps the Oxus and Jaxartes. Into the minor difficulties respecting the site of Eden it would be unprofitable to enter, and it will matter little if we accept that view, which, however, I think less probable, that it was placed in the lower part of the valley of the Euphrates. I may merely mention one particular of the Biblical description, because it throws light on the great antiquity of this geographical delineation, and has been strangely misconceived by expositors—the relation of those rivers to Cush or Ethiopia and Havilah, a tribal name derived from that of a grandson of Cush. On consulting the tenth chapter of Genesis, it will be found that the Cushites under Nimrod, very soon after the deluge, are stated to have pushed their migrations and conquests along the Tigris to the northward, and established there the first empire. It is probably this primitive Cushite empire, called Ethiopia in our translation, which in the epoch of the description of Eden occupied the Euphratean valley, and being bounded on one side by the river called Gihon, was thus believed to extend over the old site of Eden. Thus the Cush or Ethiopia of the description has no direct connection with the African Ethiopia, and speculations based on such a supposed connection are groundless. On the other hand this feature furnishes an interesting coincidence with other parts of Genesis, and throws light on many obscure points in the early history of man ; and since this Cushite empire had perished even before the time of Moses, it indicates a still more ancient tradition respecting the primeval abode of our species.

2. Before the deluge this region must have been the seat of a dense population, which, according to the Biblical account, must have made considerable advances in the arts, and at the same time sunk very low in moral debase-

ment.* Whether any remains of the central portions of this ancient population or its works exist will probably not be determined with absolute certainty till we have accurate geological
investigations of the whole country in the neighborhood of the
Caspian Sea and along the great rivers of Western Asia, though
there is nothing unreasonable in the belief that some of the
old prehistoric men whose remains are discovered in caves
and river gravels in Europe may belong to the antediluvian
race. Should such remains be found, we might infer, from
the extreme longevity and other characteristics assigned to
the antediluvians, that their skeletons would present peculiarities entitling them to be considered a well-marked variety of
the human species, and this not of a low type of physical
organization. We may also infer that the family of man very
early divided into two races—one retaining in greater purity
the moral endowments of the species, the other excelling in
the mechanical and fine arts ; and that there were rude and
savage outlying communities of men then as at present. If
the so-called palæolithic men of Europe are antediluvian, they
were probably of such outlying tribes, and possibly of the
mixed race which sprung up in the later antediluvian age,
and who are described as mighty men physically, and men of
violence. It would be quite natural that this intermixture

---

* The Bible specifies, perhaps only as the principal of these arts, music
and musical instruments by Jubal, metallurgy by Tubalcain, the domestication of cattle and the nomade life by Jabal. It is highly probable that
these inventors are introduced into the Mosaic record for a theological
reason, to point out the folly of the worship rendered to Phtha, Hephæstos, Vulcan, Horus, Phœbus, and other inventors, either traditionary representatives of the family of Lamech, or other heroes wrongly identified
with them. Very possibly their sister Naamah, "the beautiful," is introduced for the same reason, as the true original of some of the female deities
of the heathen.

of the Sethite and Cainite races should produce a race excelling both in energy and physical endowments—the "giants" that were in those days.* If any remains of the two central nations of the antediluvian period are ever discovered, we may confidently anticipate that the distinctive characteristics of these races may be detected in their osseous structures as well as in their works of art. Farther, it is to be inferred from notices in the fourth chapter of Genesis, that before the deluge there was both a nomadic and a settled population, and that the principal seat of the Cainite, or more debased yet energetic branch of the human family, was to the eastward of the site of Eden. No intimations are given by which the works of art of antediluvian times could be distinguished from those of later periods ; but that curious summary of the treasures of antediluvian man contained in the notice that the land of Havilah produced gold and agate and pearl (Gen. ii., 12) would lead us to believe that the early antediluvian age was on the whole an age of stone, in which flint for weapons, and gold and shell wampum for ornaments, were the leading kinds of wealth. On the other hand, the notices of antediluvian metallurgy, and the building and construction of the ark, would lead us to infer that the later antediluvians had attained to much perfection in some constructive arts—a conclusion which harmonizes with the otherwise inexplicable perfection of such art soon after the deluge, as evidenced not only by the story of Babel, but also by the early works of the Assyrians and Egyptians.

3. When the antediluvian population had fully proved it-

---

* I can not for a moment entertain the monstrous supposition of many expositors that the "sons of God" of these passages are angels, and the " Nephelim" hybrids between angels and men.

self unfit to enter into the divine scheme of moral renova-
tion, it was swept away by a fearful physical catastrophe.
The deluge might, in all its relations, furnish material for an
entire treatise. I may remark here, as its most important
geological peculiarity, that it was evidently a *local* convulsion.
The object, that of destroying the human race and the animal population of its peculiar centre of creation, the preservation of specimens ·of these creatures in the ark, and the
physical requirements of the case, necessitate this conclusion, which is now accepted by the best Biblical expositors,*
and which inflicts no violence on the terms of the record.
Viewed in this light, the phenomena recorded in the Bible,
in connection with geological probabilities, lead us to infer
that the physical agencies evoked by the divine power to
destroy this ungodly race were a subsidence of the region
they inhabited, so as to admit the oceanic waters, and extensive atmospherical disturbances connected with that subsidence, and perhaps with the elevation of neighboring regions.
In this case it is possible that the Caspian Sea, which is now
more than eighty feet below the level of the ocean,† and
which was probably much more extensive then than at present, received much of the drainage of the flood, and that the
mud and sand deposits of this sea and the adjoining desert
plains, once manifestly a part of its bottom, conceal any remains that exist of the antediluvian population. In connection with this, it may be remarked that, in the book of Job,
Eliphaz speaks as if the locality of those wicked nations

---

* See Lange's "Commentary on Genesis."
† The Russian surveys of 1836 made it one hundred and eight English feet ; but later authorities reduce it to eighty-three feet six inches
below the Black Sea.

which existed before the deluge was known and accessible
in his time :

> " Hast thou marked the ancient way
> Which wicked men have trodden,
> Who were seized [by the waters] in a moment,
> And whose foundations a flood swept away ?"
> —Job xxii., 15.

On comparing this statement with the answer of Job in the
26th chapter, verse 5th, it would seem that the ungodly antediluvians were supposed to be still under the waters ; a belief quite intelligible if the Caspian, which, on the latest and
most probable views of the locality of the events of this book,
was not very remote from the residence of Job,* was supposed to mark the position of the pre-Noachic population,
as the Dead Sea afterward did that of the cities of the plain.
Some of the dates assigned to the book of Job would, however, render it possible that this last catastrophe is that to
which *he* refers :

> " The *Rephaim* tremble from beneath
> The waters and their inhabitants.
> Sheol is naked before him,
> And destruction hath no covering."

The word *Rephaim* here has been variously rendered
"shades of the dead" and "giants." It is properly the
family or national name of certain tribes of gigantic Hamite men (the Anakim, Emim, etc.) inhabiting Western Asia
at a very remote period ; and it must here refer either to
them or to the still earlier antediluvian giants.†

---

* Kitto's " Bible Illustrations "—Book of Job.
† See article " Rephaim " in Kitto's " Journal of Sacred Literature."
But Gesenius and others regard it, not as an ethnic name, but as a term
for the "shades" or spirits of the dead. See Conant on Job.

It is also an important point to be noticed here that the narrative of the deluge in Genesis is given as the testimony or record of an eye-witness, and is to be so understood ; and that the terms of the record imply, not as usually held that all sorts of animals were taken into Noah's ark, but only a selection, the character of which is clearly indicated by a comparison of the five lists of animals given in the narrative. Bearing this in mind, and noticing that the writer tells of his own experience as to the rise of the water, the drifting of the ark, the disappearance of all visible shore, and the sounding fifteen cubits where a hill had before been, all the difficulties of the narrative of the deluge will at once disappear. These difficulties have in fact arisen from regarding the story as the composition of a historian, not as what it manifestly is, the log or journal of a contemporary, introduced with probably little change by the compiler of the book.

After the deluge, we find the human race settled in the plains of the Euphrates and Tigris, attracted thither by the fertility of their alluvial soils. There we find them engaging in a great political scheme, no doubt founded on recollections of the old antediluvian nationalities, and on a dread of the evils which able and aspiring men would anticipate from that wide dispersion of the human race that appears to have been intended by the Creator. in the new circumstances of the earth. They commenced accordingly the erection of a city or tower at Babel, in the plain of Shinar, to form a common bond of union, a great public work that should be a rallying-point for the race, and around which its patriotism might concentrate itself. The attempt was counteracted by an interposition of divine Providence ; and thenceforth the diffusion of the human race proceeded unchecked, carrying with it everywhere the memory of the

celebrated tower, which perpetuated itself not only in the
mounds of Assyria and Babylon and the pyramids of Egypt,
but in the teocallis and temple mounds of the New World.
The Babel enterprise is in fact the first recorded development
of that mound-building instinct which the earlier races everywhere evince, and which has been a distinguishing characteristic more especially of the Cushite or Turanian race, and has
apparently made them the teachers of constructive arts to all.
other peoples.   Perhaps a dread of the total decay and loss
of the surviving antediluvian arts in construction and other
matters may have been one impelling motive to the building
of Babel.   Perhaps it was connected with the communistic
ideas of the Turanian race, and their conflict with the patriarchal habits of the Semites.   Out of the enterprise at Babel,
however, arose a new type of evil, which, in the forms of military despotism, the spirit of conquest, hero-worship, and the
alliance of these influences with literature and the arts, has
been handed down through every succeeding age to our own
time.   The name of Nimrod, the son of Cush, has been preserved to us in the Bible, and also apparently in the tablets
and inscriptions of Assyria, as the founder of the first despotism.   This bold and ambitious man, subsequently deified
under different names, established a Hamite or Turanian
empire, which appears to have extended its sway over the
tribes occupying Southwestern Asia and Northeastern Africa,
everywhere supporting its power by force of arms, and introducing a debasing polytheistic hero-worship, and certain
forms of art probably derived from antediluvian times.   The
centre of this Cushite empire, however, gave way to the rising power of Assyria or the Ashurite branch of the sons of
Shem, at a period antecedent to the dawn of profane history,
except in its mythical form ; and when the light of secular

history first breaks upon us, we find Egypt standing forth as
the only stable representative of the arts, the systems, and
the superstitions of the old Cushite empire, of which it had
been the southern branch; while other remnants of the
Hamite races, included in the empire of Nimrod, were scattered over Western Asia, and, migrating into Europe, with or
after the ruder but less demoralized sons of Japheth, carried
with them their characteristic civilization and mythology, to
take root in new forms in Greece and Italy.*   Meanwhile
the Assyrian and Persian (Elamite) races were growing in
Middle Asia, and probably driving the more eastern remnants of the Nimrodic empire into India, borrowing at the
same time their superstitions and their claims to universal
dominion.   These views, which I believe to correspond with
the few notices in the Bible and in ancient history, and to be
daily receiving new confirmations from the investigations of
the ancient Assyrian monuments, enable us to understand
many mysterious problems in the early history of man.
They give us reason to suspect that the *principle* of the first
empire was an imitation of the antediluvian world, and that
its arts and customs were mainly derived from that source.
They show how it happens that Egypt, a country so far removed from the starting-point of man after the deluge,
should appear to be the cradle of the arts, and they account
for the Hamite and perhaps antediluvian elements, mixed

---

* On the Biblical view of this subject, the so-called Aryan mythology,
common to India and Greece, is either a derivative from the Cushite
civilization, or a spontaneous growth of the Japetic stock scattered by the
· Cushite empire.   The Semitic and Hamitic mythologies are derived from
the primeval cherubic worship of Eden, corrupted and mixed with deifica-
tion of natural objects and stages of the creative work, and with adoration
of deified ancestors and heroes.

with primeval Biblical ideas, as the cherubim, etc., in the old
heathenism of India, Assyria, and Southern Europe, and
which they share with Egypt, having derived them from the
same source.   They also show how it is that in the most remote antiquity we find two well-developed and opposite religious systems ; the pure theism of Noah, and those who retained his faith, and the idolatry of those tribes which regarded with adoring veneration the objects and stages of the
creative work, the grander powers and objects of nature, the
mighty Cainites of the world before the flood, and the postdiluvian leaders who followed them in their violence, their
cultivation of the arts, and their rebellion against God.
These heroes were identified with imaginative conceptions
of the heavenly bodies, animals, and other natural objects,
associated with the fortunes of cities and nations, with particular territories, and with war and the useful arts, transmitted under different names to one country after another,
and localized in each ; and it is only in comparatively modern times that we have been able to recognize the full certainty of the view held long since by many ingenious writers,
that among the greater gods of Egypt and Assyria, and of
consequence among those also of Greece and Rome, were
Nimrod, Ham, Ashur, Noah, Mizraim, and other worthies and
tyrants of the old world ; and to suspect that Tubalcain and
Naamah, and other antediluvian names, were similarly hon- ·
ored, though subsequently overshadowed by more recent di-
vinities.   The later Assyrian readings of Rawlinson, Hincks,
and the lamented George Smith, and the more recent works
on Egyptian antiquities, are full of pregnant hints on these
subjects.   It would, however, lead us too far from our immediate subject to enter more fully into these questions.   I
have referred to them merely to point out connecting-links

between the secular and sacred history of the earlier part of the human period, as a useful sequel to our comparison of the latter with the conclusions of science, and as furnishing hints which may guide the geologist in connecting the human with the tertiary period, and in distinguishing between the antediluvian and postdiluvian portions of the former.

It may be said, however, that all this Biblical history, however it may accord with the little that remains to us of the written annals of early Oriental nations, is entirely at variance with those modern archæological discussions which point to an immense antiquity of the human race, and to a primitive barbarism out of which all human culture was little by little evolved ; and which results of archæological investigation, while contradictory to the Hebrew Scriptures, are entirely in accord with the evolutionist philosophy. The prominence now given to such views as these renders it necessary that we should denote a special chapter to their discussion.

CHAPTER XIII.

UNITY AND ANTIQUITY OF MAN.

" These are the families of the sons of Noah, after their generations, in their nations: and by these were the nations divided in the earth after the flood."—Genesis x., 32.

THE theologians and evangelical Christians of our time, and with them the credibility of the Holy Scriptures, are supposed by many to have been impaled on a zoological and archæological dilemma, in a manner which renders nugatory all attempts to reconcile the Mosaic cosmogony with science. The Bible, as we have seen, knows but one Adam, and that Adam not a myth or an ethnic name, but a veritable man; but some naturalists and ethnologists think that they have found decisive evidence that man is not of one but of several origins. The religious tendency of this doctrine no Christian can fail to perceive. In whatever way put, or under whatever disguise, it renders the Bible history worthless, reduces us to that isolation of race from race cultivated in ancient times by the various local idolatries, and destroys the brotherhood of man and the universality of that Christian atonement which proclaims that "as in Adam all die, so in Christ shall all be made alive."

Fortunately, however, the greater weight of biological and archæological evidence is here on the side of the Bible, and philology comes in with strong corroborative proof. But just as the orthodox theologian is beginning to congratulate himself on the aid he has thus received, some of his new friends gravely tell him that, in order to maintain their view, it is necessary to believe that man has resided on earth for count- less ages, and that it is quite a mistake to suppose that his starting-point is so recent as the Mosaic deluge. Nay, some very rampant theorists of some ethnological schools try to pierce Moses and his abettors with both horns of the di- lemma at once, maintaining that men may be of different species, and yet may have existed for an enormous length of time as well. The recent prevalence of theories of evolu- tion has, however, thrown quite into the background the dis- cussions formerly active respecting the unity of man, but has, along with geological and archæological discovery, given in- creased prominence to those relating to the date of the origin of our species and the manner of its introduction.

The Bible gives us a definite epoch, that of the deluge, about 2000 to 3000 B.C., for all existing races of men; but this, according to it, was only the second starting - point of humanity, and though no family but that of Noah survived the terrible catastrophe, it would be a great error to suppose that nothing antediluvian appears in the subsequent history of man. Before the deluge there were arts and an old civiliza- tion, extending over at least two thousand years, and after the deluge men carried with them these heirlooms of the old world to commence with them new nations. This has been tacitly ignored by many of the writers who underrate the value of the Hebrew history. It may be as well for this reason to place, in a series of propositions, the principal points in Gen- esis which relate to the question of the unity of man.

1. Adam and Isha, the woman, afterward called Eve (Life- giver), in consequence of the promise of a Redeemer, com- menced a life of husbandry on their expulsion from Eden,

which, on the ordinary views of the Bible chronology, may be supposed to have occurred from 4000 to 5000 years before the Christian era; and during the lifetime of the primal pair, the sheep, at least, was domesticated. The Bible, of course, knows nothing of the imaginary continent of Lemuria, in which, according to some hypotheses, men are supposed to have had their birth from apes. A few generations after, in the time of Lamech, cattle were domesticated; and the metals copper and iron were applied to use—the latter probably meteoric iron; and hence, it may be, the Hindoo and Hellenic myths of Twachtrei and Hephæstos in connection with the thunderbolt. We learn, however, incidentally, as already mentioned, in the description of Eden in Genesis, chapter 2d, that there was a previous stone age, in which "flint, pearls or shell beads, and stream-gold" were the chief treasures of man, for this is implied in the "gold, bedolach, and onyx" of the land of Havilah. It is certain also, from the discoveries made in Assyria, on the site of Troy, and elsewhere, that the use of stone implements continued in Western Asia long after the deluge. In the time of Noah the distinction of clean and unclean beasts, and the taking of seven pairs of certain beasts and birds into the ark, imply that certain mammals and birds were domesticated.*

2. Before the flood, as already remarked, there was a division of man into two nationalities or races; and there was a citizen, an agricultural, a pastoral, and a nomadic population. Farther, the remarkable progress in the arts implied in the building of such structures as the Tower of Babel, and other temple and palace mounds in Assyria, and of the pyramids

* Genesis 4th, 5th, 6th, and 7th chapters. See also our previous remarks on the deluge.

M

of Egypt, within a few generations after the deluge, proves that a very advanced material civilization and great skill in constructive arts had been reached in antediluvian times.*

3. After the deluge, the arts of the antediluvians and their citizen life were almost immediately revived in the plain of Shinar; but the plans of the Babel leaders, like those of many others who have attempted to force distinct tribes into one nationality, failed. The guilt attributed to them probably relates to the attempt to break up the patriarchal and tribal organization, which in these early times was the outward form of true religion, in favor of some sort of national organization, not compatible with the extension of man immediately over the world, and tending to consolidation into dense communities. It may be a question here whether the tribal communism which has prevailed among the American Indians and other rude races was the primitive form of society which the Babel-builders essayed to change, or whether the Semitic patriarchal system had at first prevailed, and the Babel difficulties were connected with a conflict between this and communism or despotism, both new Turanian or Aryan introductions. In any case, Babel, and Babylon its successor, remain in the subsequent Biblical literature as types of the God-defying and antichristian systems that have succeeded each other from the time of Nimrod to this day.

4. The human race was scattered over the earth in family groups or tribes, each headed by a leading patriarch, who gave it its name. First, the three sons of Noah formed three main stems, and from these diverged several family branches. The ethnological chart in the 10th chapter of Genesis gives the principal branches under patriarchal and ethnic names;

---

* Genesis iv.

but these, of course, continued to subdivide beyond the space
and time referred to by the sacred writer. It is simply absurd
to object, as some writers have done, to the universality of
the statements in Genesis, that they do not mention in detail
the whole earth. They refer to a few generations only, and
beyond this restrict themselves to the one branch of the human family to which the Bible principally relates. We should
be thankful for so much of the leading lines of ethnological
divergence, without complaining that it is not followed out
into its minute ramifications and into all history.

5. The tripartite division in Genesis x. indicates a somewhat strict geographical separation of the three main trunks.
The regions marked out for Japheth include Europe and Northwestern Asia. The name Japheth, as well as the statements
in the table, indicate a versatile, nomadic, and colonizing disposition as characteristic of these tribes.* The Median population, the same with a portion of that now often called Aryan,†
was the only branch remaining near the original seats of the
species, and in a settled condition. The outlying portions
of the posterity of Japheth, on account of their wide dispersion,
must at a very early period have fallen into comparative bar-

---

* Japheth is "enlargement," his sons are Scythians and inhabitants of
the isles, varying in language and nationality; and Noah predicts, "God
shall enlarge Japheth, he shall dwell in the tents of Shem, Ham shall be
his servant." These are surely characteristic ethnological traits for a period so early. On the rationalist view, it may be supposed that this pre-
diction was not written until the characters in question had developed
themselves; but since the greatest enlargement of Japheth has occurred
since the discovery of America, there would be quite as good ground for
maintaining that Noah's prophecy was interpolated after the time of Columbus.

† The language of this people, the stem of the Indo-European languages,
is, though in a later form, probably that of the Aryan or Persepolitan part
of the trilingual inscriptions at Behistun and elsewhere in Persia.

barism, such as we find in historic periods all over Western and Northern Europe and Northern Asia. Owing to their habitat, the Japhetites of the Bible include none of the black races, unless certain Indian and Australian nations are outlying portions of this family. The Shemite nations showed little tendency to migrate, being grouped about the Euphrates and Tigris valleys and neighboring regions. For this reason, with the exception of certain Arab tribes, they present no instances of barbarism, and generally retained a high cerebral organization, and respectable though stationary civilization, and they possess the oldest alphabet and literature. The posterity of Ham differs remarkably from the others. It spread itself over Southern, Central, and Eastern Asia, Southern Europe, and Northern Africa, and constitutes the stock alike of the Turanian and African races, as well as probably of the American tribes. It has all along displayed a great capacity for certain forms of art and semi-civilization, but has rarely risen to the level of the Shemite and Japhetite races. It established the earliest military and monarchical institutions, and presents at the dawn of history—in Assyria, in Egypt, and India—settled and arbitrary forms in politics and religion, of a character so much resembling that of an old and corrupt civilization that we can scarcely avoid supposing that Ham and his family had preserved more than any of the other Noachian races the arts and institutions of the old world before the flood. It certainly presents itself in early postdiluvian times as the first representative and teacher of art and material civilization. The Hamite race is remarkable for the early development of pantheism and hero-worship, and for the artificial character of its culture. It presents us with the darkest colors, and in the vast solitudes of Africa and Central Asia its outlying tribes must have fallen into comparative barbarism a few centuries after the deluge. It is farther to be observed that, according to the Bible, the Canaanites and other Hamite nations spoke languages not essentially different from those of the Shemites, while the Japhetite nations were to them barbarians—"a nation whose tongue thou shalt not understand." There was, too, at the date of the dispersion of Babel, already a distinction of tongues within each of the great races of men.

6. All the divisions of the family of Noah had from the first the domesticated animals and the principal arts of life, and enjoyed these in a national capacity so soon as sufficiently numerous. The more scattered tribes, wandering into fresh regions, and adopting the life of hunters, lost the characteristics of civilization, and diverged widely from the primitive languages. We should thus have, according to the Hebrew ethnology, a central area presenting the principal stems of all the three races in a permanently civilized state. All around this area should lie aberrant and often barbarous tribes, differing most widely from the original type in the more distant regions, and in those least favorable to human health and subsistence. In these outlying regions, secondary centres of civilization might grow up, differing from that of the primitive centre, except in so far as the common principles of human nature and intercommunication might prevent this. All these conclusions, fairly deducible at once from the Mosaic ethnology and the theory of dispersion from a centre, are perfectly in accordance with observed facts, though in absolute contradiction to prevalent ethnological conclusions, based on these facts in connection with theories of development.

A multitude of Bible notices might easily be quoted illustrative of these points, and also of the consistency of the Mosaic narrative with itself. One of them may suffice here.

Abraham, who is said by the Jews to have been contemporary with Shem, as Menes by the Egyptians with Ham, at least lived sufficiently near to the time of the rise of the earliest nations to be taken as an illustration of this primitive condition of society. He was not a patriarch of the first or second rank, like Ham or Mizraim or Canaan, but a subordinate family leader several removes from the survivors of the ˙ deluge. Yet his tribe increases in comparatively few years to a considerable number. He is treated as an equal by the monarchs of Egypt and Philistia. He defeats, with a band of three or four hundred retainers, a confederacy of four Euphratean kings representing the embryo state of the Persian and Assyrian empires, and already relatively so strong that they have overrun much of Western Asia. All this bespeaks in a most consistent manner the rapid rise of many small nationalities, scattered over the better parts of wide regions, and still in a feeble condition, though inheriting from their ancestors an old civilization, and laying the foundations of powerful states. If we attach any historical value whatever to the narrative, it obviously implies that at a date of about two thousand years before Christ the regions afterward occupied by the oldest historic empires were still thinly peopled, and their dominant races little more than feeble tribes. This farther corresponds with the authentic history of all the ancient nations, however these may have been extended by previous mythical periods. About or shortly before the time of Abraham, Menes was draining for the first time the swamps of Egypt, Ninus or Nimrod was founding the Assyrian empire, the Phœnicians were founding Sidon, agriculture was being introduced into China, the Vedas were being written in India, the Persian monarchy was being founded ; and, in short, all the historical nations of the East were originating, and this apparently by springing into being with an already formed civilization.

Such being the Hebrew account of the date and early history of man, it may be proper here to compare it with such deductions from archæological and geological investigation as may seem to conflict with it, and at the same time to make some comparisons with the Turanian and Aryan traditions and speculations as to human origins. The special lines of investigation important here are: 1. Early historical records other than the Bible; 2. The diversity of human languages; 3. The geological evidence afforded by remains of prehistoric men found in caverns and other repositories. The last of these is at present that which has attained the greatest development.

1. *Early Human History.*—Had the human race everywhere preserved historical records, we should have had some certain evidence as to the places and times of origination of its tribes and peoples. Unfortunately this has not been the case. All savage and barbarous races, and many of those now civilized, have lost all records of their early history. Most of the so-called ancient nations are comparatively modern, and their history after a very short course loses itself in uncertain tradition and mythical fancies. The only really ancient nations that have given us in detail their own written history are the Hebrews, the Assyrians, the Egyptians, the Hindoos, and the Chinese. The last people, though professedly very ancient, trace their history from a period of barbarism—a view confirmed by their physical characters and the nature of their civilization; and on this account, if no other, their history can not be considered as of much archæological value. According to their own records, their earliest authentic history goes back to about 2800 B.C., and

was preceded by a prehistoric period of uncertain duration. The astronomical deductions of Schlegel, which would extend their history to 17,000 years, are evidently altogether unreliable.* The early Hindoo history is palpably fabulous or distorted, and has been variously modified and changed in comparatively modern times. There is one great and very ancient people—the Egyptian—evidently civilized from the beginning of all history, that have succeeded in transmitting to us, though only in fragments, their primeval history; and of late years constant additions have been made from inscribed tablets and monuments to our knowledge of the ancient history of the Assyrians and Chaldeans.

The Egyptian history has been gathered first from sketches by Greek travellers, and from fragments of the chronicles of Manetho, one of the later Egyptian priests; and, secondly, from the inscriptions deciphered on Egyptian monuments and papyri. It is still in a very fragmentary and uncertain state, but has been used with considerable effect to prove both the diversity of races of men and the pre-Noachic antiquity of the species. The Egyptian, in features and physical conformation, tended to the European form, just as the modern Fellahs and Berbers do; but he had a dark complexion, a somewhat elongated head and flattened lips, and certain negroid peculiarities in his limbs. His language combined many of the peculiarities of the Semitic, Aryan, and African tongues, indicating thereby great antiquity or else great intermixture, but not, as some ethnographers demand, both; most probably the former—the Egyptians being really the oldest civilized people that we certainly know, and therefore, if languages have one origin, likely to be near its root-stock.

---

* Edkins, "China's Place in Philology."

The actual history of Egypt begins from Menes, the first human king, a monarch, or rather tribal chief, who took up his abode in the flats and fens of Lower Egypt, certainly not very long after the deluge. His name has been translated "one who walks with Khem," or Ham ; one, therefore, who was contemporary with this great patriarch and god of the Egyptians, which will place his time within a few centuries of the Biblical flood. The date of Menes has been variously placed. In correction of the ordinary Hebrew chronology, we have the following attempts :

Josephus places his reign........................ 2350 B.C.
Dr. Hales' calculation............................ 2412
Manetho and the Monuments, as corrected by Syn- ⎰ 2712
   cellus and calculated by various archæologists... ⎱ to
                                              2782
Herodotus, astronomical reduction by Rennell...... 2890
Estimate by Gliddon in "Ancient Egypt".......... 2750
Bunsen, " Egypt's Place," etc.................... 4000

The truth may be somewhere near the mean of the shorter chronologies given in the list.* That of Bunsen is liable to very grave objections ; more especially as he adds to it other views, altogether unsupported by historical evidence, which would carry back the deluge to 10,000 years B.C. It rests wholly on the chronology of Manetho, who lived 300 years B.C. ; and who, even if the Egyptians then possessed authentic documents extending 3700 years before his time, may have erred in his rendering of them ; and is farther liable to grave suspicions of having merely grouped the names on the monuments of his country arbitrarily in Sothic cycles. Far-

---

* Reginald S. Poole has adduced very ingenious arguments, monumental, astronomical, and mythological, for the date B.C. 2717.

ther, they rest on an interpretation of Manetho, which supposes his early dynasties to have been successive, while good
reasons have been found to prove that many of them consist
of contemporaneous petty sovereigns of parts of Egypt. The
early parts of Manetho's lists are purely mythical, and it is
impossible to fix the point where his authentic history commences. He copied from monuments which have no consecutive dates, the precise age of which could only be vaguely
known even in his time, and which are different in their statements in different localities. It is only by making due allowance for these uncertainties that any historical value can be
attached to these earlier dynasties of Manetho. Yet Bunsen
has built on an uncertain interpretation of this writer, as
handed down in a very fragmentary and evidently garbled condition, and on the equally or more uncertain chronology of Era-
tosthenes, a system differing from all previous belief on the subject, from the Hebrew history, and from all former interpretations of the monuments and Manetho.* Discarding, there-

---

* It is curious that almost simultaneously with the appearance of Bunsen's scheme a similiar view was attempted to be maintained on geological grounds. In a series of borings in the delta of the Nile, undertaken
by Mr. Horner, there was found a piece of pottery at a depth which appeared to indicate an antiquity of 13,371 years. But the basis of the calculation is the rate of deposit (3½ inches per century) calculated for the
ground around the statue of Rameses II. at Memphis, dated at 1361 B.C.;
and Mr. Sharpe has objected that no mud could have been deposited around
that statue from its erection until the destruction of Memphis, perhaps 800
years B.C. Farther, we have to take into account the natural or artificial
changes of the river's bed, which in this very place is said to have been diverted from its course by Menes, and which near Cairo is now nearly a mile
from its former site. The liability to error and fraud in boring operations
is also very well known. It has farther been suggested that the deep
cracks which form in the soil of Egypt, and the sinking of wells in ancient
times, are other probable causes of error ; and it is stated that pieces of
burnt brick, which was not in use in Egypt until the Roman times, have

fore, in the mean time, this date, and the still older one claimed by Mariette,* we may roughly estimate the date of Menes as 2000 to 2500 years B.C.,† and proceed to state some of the facts developed by Egyptologists.

One of the most striking of these is the proof that Egypt was a new country in the days of Menes and several generations of his successors. The monuments of this period show little of the complicated idolatry, ritual, and caste system of later times, and are deficient in evidence of the refinement and variety of art afterward attained. They also show that these early monarchs were principally engaged in dyking, and otherwise reclaiming the alluvial flats; an evidence precisely of the same character with that which every traveller sees in the more recently settled districts of Canada, where the forest is giving way to the exertions of the farmer. Farther, in this primitive period, known as the "old monarchy," few domestic animals appear, and experiments seem to have been in progress to tame others, natives of the country, as the hyena, the antelope, the stork. Even the dog in the older dynasties is represented by one or at most two varieties, and the prevalent one is a wolfish-looking animal akin to the present wild or

---

been found at even greater depths than the pottery referred to by Mr. Horner. This discovery, at first sight so startling, and vouched for by a geologist of unquestioned honor and ability, is thus open to the same doubts with the Guadaloupe skeletons, the human bones in ossiferous caverns, and that found in the mud of the Mississippi ; all of which have, on examination, proved of no value as proofs of the geological antiquity of man.

* 5004 B.C.

† Perhaps the earliest certain date in Egyptian history is that of Thothmes III. of the eighteenth dynasty, ascertained by Birch on astronomical evidence as about 1445 B.C. (about 1600, Manetho); and it seems nearly certain that before the eighteenth dynasty, of which this king was the fifth sovereign, there was no settled general government over all Egypt.

half-tamed dogs of the East.* The Egyptians, too, of the earlier dynasties, are more homogeneous in their appearance than those of the later, after conquest and migration had introduced new races; and the earliest monumental notice referring to Negro tribes does not appear until the 12th dynasty, about half-way between the epoch of Menes and the Christian era, nor does any representation of the Negro features occur until, at the earliest, the 17th dynasty. This allows ample time—one thousand years at the least—for the development, under abnormal circumstances and isolation, of all the most strongly marked varieties of man. Still Egypt, even under the old monarchy, presents evidence of the continuation of antediluvian culture.†

It is obvious, in short, that the whole aspect of early Egyptian history presents to us a people already civilized taking possession of that country at a period corresponding with that of the subsidence of the Noachian deluge, and not finding there any remains of older populations. Nor have any remains of such populations been found by modern investigation. ‡

In Assyria the results of the recent discoveries, so well known through many learned· and popular works, strikingly confirm the Hebrew chronology. They indicate no slow emer-

---

* The Egyptians seem, like our modern cattle-breeders, to have taken pride in the initiation and preservation of varieties. Their sacred bull, Apis, was required to represent one of the varieties of the ox; and one can scarcely avoid believing that some of their deified ancestors must have earned their celebrity as tamers or breeders of animals. At a later period, the experiments of Jacob with Laban's flock furnish a curious instance of attempts to induce variation.

† See for evidence of these views early notices in Genesis, and Lenormant and Osburne on Egyptian Monuments and History.

‡ There is no good reason to believe the flint implements mentioned by Delanoüe and others, as found on the banks of the Nile, to be older than the historic period.

gence from barbarism, but show that in Assyria as in Egypt implements of stone and metal were used together by a primitive people, already far advanced in civilization; and the oldest historical names only carry us back to cities and sovereigns of the Abrahamic age, while the story of the primitive empire of Nimrod and the traditions of the deluge seem to have survived in more or less mythical legends. The earliest Assyrian monuments would seem to belong to a Turanian race, of which comparatively little is known, but which may correspond with the primitive Cushites of Biblical story. To these, it is true, Berosus attaches a fabulous antiquity; but this is not confirmed by the monuments. These, according to the latest facts disclosed by Smith, Rawlinson, and others, appear to fix a date of about 1800 B.C. for the foundation of the Assyrian monarchy proper, and the oldest previous date given by Assurbampal, who reigned about B.C. 668 to 626, gives 1635 years before his time, or say 2280 B.C., as the date of an Elamite king Kudarnankundi, who seems to be the leader of a primitive tribe, one of the oldest in the region, and who has been conjectured to have been the Chedorlaomer of Genesis, but was probably one of his predecessors.

We gather from the Assyrian annals that the early Turanian kings, while mound-builders like their kindred elsewhere, and acquainted with metals and with the cuneiform writing, yet constituted comparatively small nations, and were much occupied with hunting and other rude sports, and with predatory expeditions, so as to answer very nearly to the Biblical conception of the early Cushite kingdom of the valley of the Euphrates, which was probably in the same stage of culture with the nations that in a later period inhabited the valley of the Mississippi, and are known as the Alleghans.

In connection with the early history of man, much importance has been attached to the division of the early historic and prehistoric ages into the periods of Stone, Bronze, and Iron, and of the former into a Palæolithic or ancient stone age, and a more modern or Neolithic stone age. It is plain, however, that too great importance has been attached to these distinctions, and that they express rather differences of circumstances and of culture than of age, so that they have really no bearing on the Biblical chronology.

If palæolithic or rudely chipped implements are the oldest known, as they not improbably were the first tools used by man, yet their use has extended in the case of rude nations all the way up to the present time ; and in America and Northern Asia we know that their antiquity is but of yesterday, and that they were used with highly finished implements of bone, and of those softer stones that admit of being polished. No certain line can therefore be drawn even locally between a Neolithic and a Palæolithic period, especially since in localities where flint implements were extensively quarried and made, as on the banks of rivers in Northern France and Southern England, and in such places as "Grimes' Graves" and Cissbury in the latter country, where mines were sunk in the chalk for the extraction of flints, it necessarily happened that vast multitudes of unfinished or spoiled implements and weapons were left on the ground, while the better-formed specimens were for the most part taken away. This conclusion is amply supported by similar localities in America, where people well acquainted with many of the arts of life have left quantities of strictly palæolithic material. Wilson, Southall, and other writers have accumulated so many examples of this that I think the distinction of Palæolithic and Neolithic ages must now be given up by all investigators who possess ordinary judgment. A remarkable instance is the celebrated " Flint ridge " of

Ohio, which was a great quarry of flint for implements used by the ancient mound-builders, a highly civilized race, as well as by the modern Indians. Here are found countless multitudes of palæolithic flint implements of all the ordinary types, but which are merely the unfinished material of workers capable of producing the most exquisite implements. There can be scarcely a doubt that the palæolithic implements of the European gravels, in so far as they are the workmanship of man, are in like manner merely the relics of old flint quarries.*

Possibly a more accurate measurement of time for particular regions of the world might be deduced from the introduction of bronze and iron. If the former was, as many antiquarians suppose, a local discovery in Europe, and not introduced from abroad, it can give no measurement of time whatever. In America, as the facts detailed by Dr. Wilson show, while a bronze age existed in Peru, it was the copper age in the Mississippi Valley, and the stone age elsewhere; and these conditions might have co-existed for any length of time, and could give no indication of relative dates. On the other hand, the iron introduced by European commerce spread at once over the continent, and came into use in the most remote tribes, and its introduction into America clearly marks an historical epoch. With regard to bronze in Europe, we must bear in mind that tin was to be procured only in England and Spain, and in the latter in very small quantity; the mines of Saxony do not seem to have been known till the Middle Ages. We must further consider that tin ore is a substance not metallic in appearance, and little likely to attract the attention of savages; and that, as we gather from a hint of Pliny, it was probably first observed, in the West at least, as stream tin, in

---

* Wilson, " Prehistoric Man," 2d edition, p. 68.

the Spanish gold washings. Lastly, when we place in connection with these considerations the fact that in the earliest
times of which we have certain knowledge, the tin trade of
Spain and England was monopolized by the Phœnicians, there
seems to be a strong probability that the extension of the trade
of this nation to the western Mediterranean really inaugurated
the bronze period. The only valid argument against this is the
fact that moulds and other indications of native bronze cast-
ing have been found in Switzerland, Denmark, and elsewhere;
but these show nothing more than that the natives could re-
cast bronze articles, just as the American Indians can forge
fish-hooks and knives out of nails and iron hoops. Other
considerations might be adduced in proof of this view, but
our limits will not permit us to refer to them. The important
questions still remain : When was this trade commenced, and
how rapidly did it extend itself from the sea-coast across Europe? The British tin trade must have been in existence in
the time of Herodotus, though his notion of the locality was
not more definite than that it was in the extremity of the
earth. The Phœnician settlements in the western Mediterranean must have existed as early as the time of Solomon,
when "ships of Tarshish" was the general designation of sea-
going ships for long voyages. How long previously these col-
onies existed we do not know; but considering the great scar-
city and value of tin in those very ancient times, we may infer
that perhaps only the Spanish, and not the British deposits were
known thus early; or that the Phœnicians had only indirect
access to the latter. Perhaps we may fix the time when these
traders were able to supply the nations of Europe with abundance of bronze in exchange for their products, at, say 1000 to
1200 B.C., as the earliest probable period ; and possibly from
one to two centuries would be a sufficient allowance for the

complete penetration of the trade throughout Europe. But
of course wars or migrations might retard or accelerate the
process; and there may have been isolated spots in which a
partial stone period extended up to those comparatively recent times in which first the Greek trade, and afterward the
entire overthrow of the Carthaginian power by the Romans,
terminated forever the age of bronze and substituted the age
of iron. This would leave, according to our ordinary chronologies, at least ten or fifteen centuries for the postdiluvian
stone period in Europe and Western Asia, a time quite sufficient in our view for all that part of it represented by such
monuments as the Danish shell-heaps or the platform habi-
tations of the Swiss lakes; leaving the remains of the prehistoric caverns and river gravels for the antediluvian period.
A few facts in illustration of these points, and also of the
Biblical history, may be mentioned here.

We know perfectly that the early Chaldeans of the Euphratean valley were acquainted with the use of metals—bronze
certainly, and at a very early date iron; yet flint knives and
other implements of stone are found under circumstances
which show that they were used in the palmy days of the
Assyrian empire. The inhabitants of Egypt were acquainted
with bronze and iron long before the date of the Exodus, yet
the Egyptians used stone knives for some purposes up to a
comparatively modern time. Joshua used stone knives for
the purpose of circumcision; and according to Herodotus
there were Ethiopians in the army of Xerxes who used stone-
tipped arrows. If any antiquarian were to stumble on the
"hill of the foreskins"—a mound under which were buried in
all probability the multitudinous flint flakes used in the circumcision of the thousands of Israel—or the grave in which
some of the Ethiopian auxiliaries of Xerxes were buried with

their flint arrow-heads and javelins of antelopes' horn, how absurd would be the inference that these repositories were of the palæolithic age. Nay, so late as 1870 a traveller was informed that the Bagos, a people of Abyssinia, still made and used stone hatchets and flint knives.*

In Europe we find reason to believe that the Ligurians of Northwestern Italy were flint-folk of very rude type until they were conquered by the Gauls about 400 B.C.† Though the Gauls, Britons, and Germans of the age of Julius Cæsar had iron weapons, yet it is evident that the metal was very scarce, and that bronze was more common ; and in confirmation of this it is found that in the trenches before Alize, the Alesia of Cæsar, where the final struggle of the Roman general with Vercingetorix took place, weapons of stone, bronze, and iron are intermixed. All over the more northern parts of Europe there is the best reason to believe that the use of stone and bronze continued to a much later period, and locally until long after the Christian era. It is clear that such facts as these must greatly modify our ideas of the probable age of the Swiss lake villages, and should induce the greatest caution in claiming any special antiquity for particular classes of implements.

One of the most remarkable discoveries of modern times is that of the site of ancient Troy by Dr. Schliemann, and it affords clear and decisive evidence as to the historic value of the ages to which we have referred.

Troy was destroyed by the Greeks perhaps about 1300 B.C., and we know from Homer that this was in what for the

* Southall has accumulated a great number of these facts in his book on the antiquity of man.
† Professor Issel, quoted in *Popular Science Monthly.*

Greeks and Trojans may properly be termed the copper age, weapons and armor of that metal being in common use, and also the mode of burial by cremation. We may well suppose that at that early date the stone age was still in full force in Northern Europe and Asia, and in the mountains of Switzerland; and as the tin mines of England had not yet been reached, bronze was scarce and dear even in Eastern Europe and Asia. Now Schliemann has disinterred the undoubted Trojan Ilium on the hill of Hissarlik; but he finds it to be only one of several buried cities, and the succession of strata will be most clearly seen in the section on the following page, compiled from his clear and circumstantial descriptions. It is needless to say that this presents a succession of the stone age to one of comparatively high civilization. It also forms an epitome of that of the whole East, and of primitive man in general, in some very important respects. We have first, at a date probably coeval with that of the earliest monarchies of Assyria and Egypt, a primitive people whose arts and mode of life remind us strongly of the American Toltecans and Peruvians.* Schliemann supposes them to have been Aryan, but they were more probably of Turanian race. They must have occupied the site for a very long time. They were succeeded by a more cultivated people of fine physical organization, yet possibly still Turanians or primitive Aryans, who by trade or plunder had accumulated large stores of metallic wealth, and had made advances in the arts of life placing them on a level with the early Phœnicians and Egyptians, with whom they probably had intercourse. These

---

* Wilson has remarked the striking similarity of the pottery of these people to American fictile wares. This similarity applies also to the early Cyprian art.

Surface. Fifth stratum to 6½ feet........	The Greek Ilium, with buildings and objects of art characteristic of the Hellenic civilization of historic periods.
Fourth stratum to 13 feet......	A second barbarous people, but probably allied to the first. Very coarse pottery. Implements and weapons of copper or bronze—stone knives and saws.
Third stratum to 23 feet........	Barbarous people occupying the site of Troy. Rude stone implements and rude pottery. Buildings of small stones and clay. Some objects of pottery found here would on American sites be regarded as probably tobacco-pipes.
Second stratum to 33 feet......	Homeric Troy. Implements and weapons of copper, bronze, and stone. Pottery, some of it of Peruvian and ancient Cypriot types. Fine gold jewelry, and gold and silver vessels. Armor similar to that described by Homer. Stone buildings and walls. This city had been sacked and burned.
First stratum to 46 or 53 feet... Rock.	Primitive or prehistoric Troy. Stone implements, polished and chipped. Millstones, copper nails, pottery—some with patterns curiously resembling those of America—bone implements, terra-cotta disks. Stone buildings.

were the Trojans of the Homeric poems, and the destruction of their city was probably in the first instance celebrated in their own native songs, which Homer at a date but little later[*] wove into his magnificent poem, and idealized and exaggerated. The Trojans worshipped an owl-headed goddess—the Athena

[*] I agree with Gladstone's conclusions as to the date and country of Homer.

of the Homeric poems ; and from symbols found are believed also to have had the worship of a sacred tree, and of fire or of the Sun. All of these are widespread superstitions over both the Old and New World. But while Troy flourished · there were barbarous nations not far off still in the stone age ; and when the city had fallen, these, possibly in successive hordes, took possession of the fertile plain and used the old city as their stronghold, perhaps till the foundation of the Greek city about 650 B.C. I have sketched in some detail these interesting discoveries, as they so clearly illustrate an actual succession of ages, and so conclusively show the uncertainty of the classification into ages of stone and metal, except when taken in connection with the precise circumstances of each locality.

I have referred above only to the question of historic or postdiluvian man. We have still to consider what remains exist of antediluvian man. These may be studied in connection with our third head of geological evidences of man's antiquity ; for if the Mosaic narrative be true, the diluvial catastrophe must have constituted a physical separation between historic man and prehistoric; since, in so far as antediluvian ages are concerned, all are prehistoric or mythical everywhere except in the sacred history itself. Antediluvian men may thus in geology be Pleistocene as distinguished from modern, or Palæocosmic as distinguished from Neocosmic.*

2. *Language in Relation to the Antiquity of Man.*—In many animals the voice has a distinctive character ; but in man it has an importance altogether peculiar. The gift of speech is

---

* I suggested these terms in my lectures published under the title " Nature and the Bible," 1875.

one of his sole prerogatives, and identity in its mode of exercise is not only the strongest proof of similarity of psychical constitution, but more than any other character marks identity of origin. The tongues of men are many and various; and at first sight this diversity may, as indeed it often does, convey the impression of radical diversity of race. But modern philological investigations have shown many and unexpected links of connection in vocabulary or grammatical structure, or both, between languages apparently the most dissimilar. I do not here refer to the vague and fanciful parallels with which our ancestors were often amused, but to the results of sober and scientific inquiry. "Nothing," says Professor Max Müller, "necessitates the admission of different independent beginnings for the material elements of the Turanian, Semitic, and Aryan branches of speech; nay, it is possible even now to point out radicals which, under various changes and disguises, have been current in these three branches ever since their first separation." Of the truth of this I have convinced myself by some original investigation, and also of the farther truth that of this radical unity of all human tongues there is more full evidence than many philologists are disposed to admit, and that the results of future study must be to connect more and more with each other the several main stems of language. Whether this results merely from the psychical unity of the human race, or from the historical derivation of languages from one root, is not so material as the fact of unity; but that the latter is implied it would not be difficult to show.* Let us examine for a little

---

* Since these words were written I have read the remarkable book of Edkins on the Chinese language, which supplies much additional information.

these results as they are presented to us by Latham, Müller, Bunsen, and other modern philologists.

A convenient starting-point is afforded by the great group of languages known as the Indo-European, Japhetic, or Aryan. From the Ganges to the west coast of Ireland, through Indian, Persian, Greek, Italian, German, Celt, runs one great language—the Sanscrit and the dark Hindoo at one extreme, the Erse and the xanthous Celt at the other. No one now doubts the affinity of this great belt of languages. No one can pretend that any one of these nations learned its language from another. They are all decided branches of a common stock. Lying in and near this area are other nations—as the Arabs, the Syrians, the Jews—speaking languages differing in words and structure—the Semitic tongues. Do these mark a different origin? The philologists answer in the negative, pointing to the features of resemblance which still remain, and above all to certain intermediate tongues of so high antiquity that they are rather to be regarded as root-stocks from which other languages diverged than as mixtures. The principal of these is the ancient Egyptian, represented by the inscriptions on the monuments of that wonderful people, and by the more modern Coptic, which, according to Bunsen and Latham, presents decided affinities to both the great classes previously mentioned, and may be regarded as strictly intermediate in its character. It has accordingly been designated by the term Sub-Semitic.* But it shares this character with all or nearly all the other African languages, which bear strong marks of affinity to the

---

* Donaldson has pointed out (British Association Proceedings, 1851) links of connection between the Slavonian or Sarmatian tongues and the Semitic languages, which in like manner indicate the primitive union of the two great branches of languages.

Egyptian and Semitic tongues. On this subject Dr. Latham
says, "That the uniformity of languages throughout Africa is
greater than it is either in Asia or in Europe, is a statement
to which I have not the least hesitation in committing myself."* To the north the Indo-European area is bounded by
a. great group of semi-barbarous populations, mostly with
Mongolian features, and speaking languages which have been
grouped as Turanian. These Turanian languages, on the
one hand, graduate without any break into those of the Es-
quimaux and American Indians; on the other, according to
Müller and Latham, they are united, though less distinctly,
with the Semitic and Japhetic tongues. They not improbably represent in more or less altered forms the most primitive stock of language from which both the Semitic and
Japhetic groups have branched. Another great area on the
coasts and in the islands of the Pacific is overspread by the
Malay, which, through the populations of Transgangetic India, connects itself with the great Indo-European line. Mr.
Edkins, in his remarkable book on "China's Place in Philology," has collected a large amount of fact tending to show
that the early Chinese in its monosyllabic radicals presents
root-forms traceable into all the stocks of human speech in
the Old World ; and the American languages would have furnished him with similar lines of affinity. If we regard physical characters, manners, and customs, and mythologies, as
well as mere language, it is much easier thus to link together nearly all the populations of the globe. In investigations
of this kind, it is true, the links of connection are often delicate and evanescent; yet they have conveyed to the ablest

---

* "Man and his Migrations." See also "Descriptive Ethnology,"
where the Semitic affinities are very strongly brought out.

investigators the strong impression that the phenomena are rather those of division of a radical language than of union of several radically distinct.

This impression is farther strengthened when we regard several results incidental to these researches. Latham has shown that the languages of men may be regarded as arranged in lines of divergence, the extreme points of which are Fuego, Tasmania, Easter Island ; and that from all these points they converge to a common centre in Western Asia, where we find a cluster of the most ancient and perfect languages; and even Haeckel is obliged to adopt in his map of the affiliation of races of men a similar scheme, though he, without any good historical or scientific evidence, extends it back into the imaginary lost continent of Lemuria. Farther, the languages of the various populations differ in proceeding from these centres in a manner pointing to degeneracy such as is likely to occur in small and rude tribes separating from a parent stock. These lines of radiation follow the most easy and probable lines of migration of the human race spreading from one centre. It must also be observed that in the primary migration of men, there must of necessity have been at its extreme limits outlying and isolated tribes, placed in circumstances in which language would very rapidly change ; especially as these tribes, migrating or driven forward, would be continually arriving at new regions presenting new circumstances and objects. When at length the utmost limit in any direction was reached, the inroads of new races of population would press into close contact these various tribes with their different dialects. Where the distance was greatest before reaching this limit, we might expect, as in America, to find the greatest mutual variety and amount of difference from the original stock. After the

N

primary migration had terminated, the displacements arising
from secondary migrations and conquests, would necessarily
complicate the matter by breaking up the original gradations
of difference, and thereby rendering lines of migration difficult to trace.

Taking all these points into the account, along with the
known tendencies of languages in all circumstances to vary,
it is really wonderful that philology is still able to give so decided indications of unity.

There is, in the usual manner of speaking of these subjects,
a source of misapprehension, which deserves special mention
in this place. The Hebrew Scriptures derive all the nations
of the ancient world from three patriarchs, and the names of
these have often been attached to particular races of men
and their languages; but it should never be supposed that
these classifications are likely to agree with the Bible affiliation. They may to a certain extent do so, but not necessarily or even probably. In the nature of the case, those portions of these families which remained near the original centre, and in a civilized state, would retain the original language
and features comparatively unchanged. Those which wandered far, fell into barbarism, or became subjected to extreme
climatic influences, would vary more in all respects. Hence
any general classification, whether on physical or philological
characters, will be likely to unite, as in the Caucasian group
of Cuvier, men of all the three primitive families, while it will
separate the outlying and aberrant portions from their main
stems of affiliation. Want of attention to this point has led to
much misconception; and perhaps it would be well to abandon altogether terms founded on the names of the sons of
Noah, except where historical affiliation is the point in question. It would be well if it were understood that when the

terms Semitic, Japhetic,* and Hametic are used, direct reference is made to the Hebrew ethnology; and that, where
other arrangements are adopted, other terms should be used.
It is obviously unfair to apply the terms of Moses in a different way from that in which he uses them. A very prevalent error of this kind has been to apply the term Japhetic
to a number of nations not of such origin according to the
Bible; and another of more modern date is to extend the
term Semitic to all the races descended from Ham, because
of resemblance of language. It should be borne in mind
that, assuming the truth of the Scriptural affiliation, there
should be a "central" group of races and languages where
the whole of the three families meet, and "sporadic"† groups
representing the changes of the outlying and barbarous
tribes.

While, however, all the more eminent philologists adhere
to the original unity of language, they are by no means agreed
as to the antiquity of man; and some, as for instance Latham
and Dr. Max Müller, are disposed to claim an antiquity for
our species far beyond that usually admitted. In so far as
this affects the Bible history, it is important, inasmuch as this
would appear to limit the possible antiquity of all languages
to the time of the deluge. The date of this event has been
variously estimated, on Biblical grounds, at from 1650 B.C.
(Usher) to 3155 B.C. (Josephus and Hales); but the longest
of these dates does not appear to satisfy the demands of philology. The reason of this demand is the supposed length

---

* I can scarcely except such terms as "Japetic" and "Japetidæ," for
Iapetus can hardly be any thing else than a traditional name borrowed
from Semitic ethnology, or handed down from the Japhetic progenitors
of the Greeks.

† See art. "Philology," Encyc. Brit.

of time required to effect the necessary changes. The subject is one on which definite data can scarcely be obtained. Languages change now, even when reduced to a comparatively stable form by writing. They change more rapidly when men migrate into new climates, and are placed in contact with new objects. The English, the Dutch, and the German were perhaps all at the dawn of the mediæval era Mæso-Gothic. At the same rate of change, allowing for greater barbarism and greater migrations, they may very well have been something not far from Egyptian or Sanscrit 2000 years before Christ. The truth is that present rates of variation afford no criterion for the changes that must occur in the languages of small and isolated tribes lapsing into or rising from barbarism, possessing few words, and constantly requiring to name new objects ; and until some ratio shall have been established between these conditions and those of modern languages, fixed by literature and by a comparatively stationary state of society, it is useless to make any demands for longer time on this ground.*

Even in the present day, Moffat informs us that in South Africa the separation of parts of a tribe, for even a few months, may produce a notable difference of dialect. If we take the existing languages of civilized men whose history is known, we shall find that it is impossible to trace many of them back as far as the Christian era, and when we have passed over even half that interval, they become so different as to be unintelligible to those who now speak them. Where there

---

* Grammatical structure is no doubt more permanent than vocabulary, yet we find great changes in the former, both in tracing cognate languages from one region to another, and from period to period. The Indo-Germanic languages in Europe furnish enough of familiar instances.

are exceptions to this, they arise entirely from the effects
of literature and artificial culture. While, therefore, there is
good ground in philology for the belief in one primitive language, there seems no absolute necessity to have recourse
even to the confusion of tongues at Babel to explain the diversities of language.* Farther, the Bible carries back the
Semitic group of languages at least to the time of the Deluge,
but it does not seem necessary on the mere ground of antediluvian names, to carry it any farther back, and the Assyrian
inscriptions show the co-existence of Turanian and Semitic
tongues at the dawn of history in the region of the Euphrates
and Tigris. One or other of these—or a monosyllabic language underlying it—was probably an antediluvian tongue,
and the other a very early derivative ; and both history and
philology would assign the precedence to the Turanian language, which was probably most akin to that which had descended from antediluvian times, and which at that early
period of dispersion indicated in the Bible story of Babel, had
begun to throw off its two great branches of the Aryan and
Semitic languages. These, proceeding in two dissimilar lines
of development, continue to exist to this day along with the
surviving portions of the uncultivated Turanian speech. We
may thus assume, as most in accordance with biblical history
and philological probability, a primitive Turanian tongue, a
somewhat rapid divergence of Semitic and Aryan branches,
and a slower development of these—the whole within the
time of postdiluvian history.

---

* It is fair, however, to observe that the Bible refers the first great divergence of language to a divine intervention at the Tower of Babel. The
precise nature of this we do not know ; but it would tend to diminish the
time required.

CHAPTER XIV.

UNITY AND ANTIQUITY OF MAN—*(Continued.)*

" By the word of God the heavens were from of old, and the earth, formed out of water, and by means of water, by which waters the world that
then was, being overflowed with water, perished."—2 Peter iii., 5, 6.

3. *Geological Evidence as to the Antiquity of Man.*—No geological fact can now be more firmly established than the ascending progression of animal life, whereby from the early invertebrates of the Eozoic and Primordial series we pass upward
through the dynasties of fishes and reptiles and brute mammals to the reign of man. In this great series man is obviously the last term ; and when we inquire at what point he
was introduced, the answer must be in the later part of the
great Cainozoic or Tertiary period, which is the latest of the
whole. Not only have we the negative fact of the absence
of his remains from all the earlier Tertiary formations, but
the positive fact that all the mammalia of these earlier ages
are now extinct, and that man could not have survived the
changes of condition which destroyed them and introduced
the species now our contemporaries. This fact is altogether
independent of any question as to the introduction of species
by derivation or by creation. The oldest geological period
in which any animals nearly related in structure to man
occur is that named the Miocene, and no traces of man
have as yet been found in any deposits of this age. All human
remains known belong either to the Pleistocene or Modern.

Now the Pleistocene was characterized by one of those periods of glacial cold which have swept over the earth—by one of those great winters which have so chilled the continents that few forms of life could survive them—and man comes in at the close of this cold period, in what is called the Post-glacial age. Some geologists, it is true, hold to an interglacial warm period, in which man is supposed to have existed,.but the evidence of this is extremely slender and doubtful, and it carries back in any case human antiquity but a very little way. I have, in my "Story of the Earth and Man," shown reason for the belief, in which I find Professor Hughes, of Cambridge, coincides with me,* that the interglacial periods are merely an ingenious expedient to get rid of the difficulties attending the hypothesis of the universal glaciation of the northern hemisphere.

But, though man is thus geologically modern, it is held that historically his existence on earth may have been very ancient, extending perhaps ten or twenty, or even a hundred times longer than the period of six or seven thousand years supposed to be proved by sacred history. Let us first, as plainly and simply as possible, present the facts supposed thus to extend the antiquity of man, and then inquire as to their validity and force as arguments in this direction.

The arguments from geology in favor of a great antiquity for man may be summarized thus: (1) Human remains are found in caverns under very thick stalagmitic crusts, and in deposits of earth which must have accumulated before these stalagmites began to form, and when the caverns were differently situated with reference to the local drainages. (2) Remains of man are found under peat-bogs which have grown

---

* Lecture in the Royal Institution, March 24, 1876.

so little in modern times that their antiquity on the whole must be very great.   (3) Implements, presumably made by men, are found in river-gravels so high above existing river-beds that great physical changes must have occurred since they were accumulated.   (4) One case is on record where a human bone is believed to have been found under a deposit of glacial age.   (5) Human remains have been found under circumstances which indicate that very important changes of level have taken place since their accumulation.   (6) Human remains have been found under circumstances which indicate great changes of climate as intervening between their date and that of the modern period.   (7) Man is known to have existed, in Europe at least, at the same time with some quadrupeds formerly supposed to have been extinct before his introduction.   (8) The implements, weapons, etc., found in the oldest of these repositories are different from those known to have been used in historic times.

These several heads include, I think, all the really material evidence of a geological character.   It is evidence of a kind not easily reducible into definite dates, but there can be no doubt that its nature, and the rapid accumulation of facts within a small number of years, have created a deep and widespread conviction among geologists and archæologists that we must relegate the origin of man to a much more remote antiquity than that sanctioned by history or by the Biblical chronology.   I shall first review the character of this evidence, and then state a number of geological facts which bear in the other direction, and have been somewhat lost sight of in recent discussions.   Of the facts above referred to, the most important are those which relate to caverns, peat-bogs, and river-gravels.   We may, therefore, first consider the nature and amount of this evidence.

That the reader may more distinctly understand the geological history of these more recent periods of the earth's history which are supposed to have witnessed the advent of man, in Western Europe at least, I quote the following summary from Sir Charles Lyell of the more modern changes in that portion of the world. These are :

" First, a continental period, toward the close of which the forest of Cromer flourished : when the land was at least 500 feet above its present level, perhaps much higher. * * * The remains of *Hippopotamus major* and *Rhinoceros etruscus*, found in beds of this period, seem to indicate a climate somewhat milder than that now prevailing in Great Britain. [This was a *Preglacial* era, and may be regarded as belonging to the close of the Pliocene tertiary.]

" Secondly, a period of submergence, by which the land north of the Thames and Bristol Channel, and that of Ireland, was generally reduced to * * * an archipelago. * * * This was the period of great submergence and of floating ice, when the Scandinavian flora, which occupied the lower grounds during the first continental period, may have obtained exclusive possession of the only lands not covered with perpetual snow. [This represents the Glacial period ; but according to the more extreme glacialists only a portion of that period.]

" Thirdly, a second continental period, when the bed of the glacial sea, with its marine shells and erratic blocks, was laid dry, and when the quantity of land equalled that of the first period. * * * During this period there were glaciers in the higher mountains of Scotland and Wales, and the Welsh glaciers * * * pushed before them and cleared out the marine drift with which some valleys had been filled during the period of submergence. * * * During this last period

N 2

the passage of the Germanic flora into the British area took place, and the Scandinavian plants, together with northern insects, birds, and quadrupeds, retreated into the higher grounds. * * *

" Fourthly, the next and last change comprised the breaking up of the land of the British area once more into numerous islands, ending in the present geographical condition of things. There were probably many oscillations of level during this last conversion of continuous land into islands, and such movements in opposite directions would account for the occurrence of marine shells at moderate heights above the level of the sea, notwithstanding a general lowering of the land. * * * During this period a gradual amelioration of temperature took place, from the cold of the glacial period to the climate of historical times."*

The second continental period above referred to is that which appears on the best evidence to have been the time of the introduction of man ; but such facts as that of the Settle Cave, and the implements of the breccia in Kent's Cave, if rightly interpreted, would make man preglacial or " interglacial."

The deposits found in caverns in France, Switzerland, Germany, Belgium, and England have afforded a large proportion of the remains from which we derive our notions of the most ancient prehistoric men of Europe. From the Belgian caves, as explored by M. Dupont, we learn that there were two successive prehistoric races, both rude or comparatively uncivilized. The first were men of Turanian type, but of great bodily stature and high cerebral organization, and showing remarkable skill in the manufacture of imple-

* " Antiquity of Man," 4th ed.

ments and ornaments of bone and ivory. These men are believed to have been contemporary with the earlier postglacial mammals, as the mammoth and hairy rhinoceros, and to have lived at a time when the European land was more extensive than at present, stretching far to the west of Ireland, and connecting Great Britain with the Continent. The skeletons found at Cro-Magnon, Mentone, and elsewhere in France fully confirm the deductions of Dupont as to this earliest race of Palæocosmic, Palæolithic, or antediluvian man. This grand race seems to have perished or been driven from Europe by the great depression of the level of the land which inaugurated the modern era, and which was probably accompanied by many oscillations of level as well as by considerable changes of climate. They were succeeded by a second race, equally Turanian in type, but of small stature, and resembling the modern Lapps. These were the "allophylian" peoples displaced by the historical Celts, and up to their time the reindeer seems to have existed abundantly in France and Germany. These two successive prehistoric populations have been termed respectively men of the "mammoth" age and men of the "reindeer" age. The Bible record would lead us to regard the earlier and gigantic men as antediluvian, and the smaller or Lappish race as postdiluvian. We may therefore, having already at some length considered the postdiluvian age, take up the mode of occurrence of the remains of the earlier of the two races—that of the mammoth age.

The caverns themselves may be divided into those of residence, of sepulture, and of driftage, though one cavern has often successively assumed two at least of these characters. In the caverns of residence large accumulations have been formed of ashes, charcoal, bones, and other débris of cookery, among which are found flint and bone implements, the gen-

eral character of which, as well as that of the needles, stone
hammers, mortars for paint, and other domestic appliances,
are not more dissimilar from those of the Red Indian and
Esquimau races in North America than these are from one
another, and in many things, as in the bone harpoons, the resemblance is very striking indeed.   In tendency to imitative
art, and in the skill of their delineations of animals, the prehistoric men seem to have surpassed all the American races
except the semi-civilized mound-builders and the more cultivated Mexican and Peruvian nations.   With regard to the
residence of these men of the mammoth age in caverns, several things are indicated by American analogies to which
some attention should be paid.

It is not likely that caverns were the usual places of residence of the whole population.   They may have been winter
houses for small tribes and detached families of fugitives or
outlaws, or they may have been places of resort for hunting
parties at certain seasons of the year.   The large quantities
of broken and uncooked bones of particular species, as of the
horse and reindeer, in some of the caverns, would farther indicate a habit of making great battues, like those of the
American hunting-tribes, at certain seasons, and of preparing
quantities of pemmican or dried meat preserved with marrow
and fat for future use.   The number of bone needles found
in some of the caves would seem to hint that, like the Americans, they sewed up their pemmican in skin bags.   The multitude of flint flakes and of rude stone implements applicable
to breaking bones certainly indicates a wholesale cutting of
flesh and preparation of marrow.   In the " Story of the
Earth," I have suggested in connection with this that there
may have been towns or villages of these people unknown
to us, and which would afford higher conceptions of their

progress in the arts. This anticipation appears recently to have been realized in the discovery of such a town or fortified village of the mammoth age at Soloutre, in France, and which seems to afford evidence that these ancient people had already domesticated the horse, using it as food as well as a beast of burden, in the manner of the Khirgis and certain other Tartar tribes of Central Asia.* This, with the undoubtedly high cerebral organization indicated by the skulls of the mammoth age, notably raises our estimate of the position of man at this early date.

With regard to caves of sepulture, the same remark may be made as with regard to the caves of residence. They do not seem to have been the burial-places of large populations, but only occasional places of interment, few bodies being found in them, and these often interred in the midst of culinary débris, evidencing previous or contemporary residence. With regard to the latter, it seems to have been no uncommon practice with some North American tribes to bury the dead either in the floors of their huts or in their immediate proximity. It is probable, however, that the few examples known of caves of sepulture of this period indicate not tribal or national places of burial, but occasional and accidental cases, happening to hunting or war parties, perhaps remote from their ordinary places of residence. In so far as method of burial is concerned, the men of the Palæocosmic or Mammoth age seem to have buried the dead extended at full length, and not in the crouching posture usual with some later races. Like the Americans, they painted the dead man, and buried him with his robes and ornaments, and probably with his weapons, thus intimating their belief in happy hunting-

---

* Southall, *Op. cit.*

grounds beyond the grave.* I may remark here that all the known interments of the mammoth age indicate a race of men of great cerebral capacity, with long heads and coarsely marked features, of large stature and muscular vigor, surpassing indeed much in all these respects the average man of modern Europe. These characteristics befit men who had to contend with the mammoth and his contemporaries, and to subdue the then vast wildernesses of the eastern continent, and they correspond with the Biblical characteristics of antediluvian man.

Among caves of driftage may be classed some of those near Liège, in Belgium, and, partially at least, those of Kent's Hole and Brixham, in England. In these only disarticulated remnants of human skeletons, or more frequently only flint implements, some of them of doubtful character, have been found. In my "Story of the Earth," I have taken the carefully explored Kent's Cavern of Torquay as a typical example, and have condensed its phenomena as described by Mr. Pengelly. I now repeat this description, with some important emendations suggested by that gentleman in more recent reports and in private correspondence.

The somewhat extensive and ramifying cavern of Kent's Hole is an irregular excavation, evidently due partly to fissures or joints in limestone rock, and partly to the erosive action of water enlarging such fissures into chambers and galleries. At what time it was originally cut we do not know, but it must have existed as a cavern at the close of the Pliocene or beginning of the Post-pliocene period, since which time it has been receiving a series of deposits which have quite filled up some of its smaller branches.

---

* The Mentone skeleton described by Dr. Rivière gives evidence of these facts.

First and lowest, according to Mr. Pengelly, of the deposits as yet known, is a "breccia," or mass of broken and rounded stones, with hardened red clay filling the interstices. Some of the stones are of the rock which forms the roof and walls of the cave, but the greater number, especially the rounded ones, are from more distant parts of the surrounding country. Many are fragments of grit from the Devonian beds of adjacent hills. There are also fragments of stalagmite from an old crust broken up when the breccia was deposited, and possibly belonging to Pliocene times. In this mass, the depth of which is unknown, are numerous bones, nearly all of one kind of animal, the cave bear or bears, for there may be more than one species—creatures which seem to have lived in Western Europe from the close of the Pliocene down to the modern period. They must have been among the earliest and most permanent tenants of Kent's Hole at a time when its lower chambers were still filled with water. Teeth of a lion and of the common fox also occur in this deposit, but rarely. Next above the breccia is a floor of "stalagmite," or stony carbonate of lime, deposited from the drippings of the roof, and in some places more than twelve feet thick. This also contains bones of the cave bear, deposited when there was less access of water to the cavern. Mr. Pengelly infers the existence of man at this time from the occurrence of chipped flints supposed to be artificial ; but which, in so far as I can judge from the specimens described and figured, must still be regarded as of doubtful origin.

After the old stalagmite floor above mentioned was formed, the cave again received deposits of muddy water and stones ; but now a change occurs in the remains embedded. This stony clay, or "cave earth," has yielded an immense quantity of teeth and bones, including those of the elephant, rhinoce-

ros, horse, hyena, cave bear, reindeer, and Irish elk. With
these were found weapons of chipped flint, and harpoons,
needles, and bodkins of bone, precisely similar to those of
the North American Indians and other rude races. The
"cave earth" is four feet or more in thickness. It is not
stratified, and contains many fallen fragments of rock, rounded stones, and broken pieces of stalagmite. It also has
patches of the excrement of hyenas, which the explorers suppose to indicate the temporary residence of these animals;
and besides fragments of charcoal scattered in the mass,
there is in one spot, near the top, a limited layer of burned
wood, with remains which indicate the cooking and eating of
repasts of animal food by man. It is clear that when this
bed was formed the cavern was liable to be inundated with
muddy water, carrying stones and perhaps some of the
bones and implements, and breaking up in places the old
stalagmite floor.* One of the most puzzling features, especially to those who take an exclusively uniformitarian
view, is that the entrance of water-borne mud and stones
implies a level of the bottom of the water in the neighboring
valleys of nearly one hundred feet above its present height.
The cave earth is covered by a second crust of stalagmite,
less dense and thick than that below, and containing only
a few bones, which are of the same general character with
those beneath, but include a fragment of a human jaw with
teeth. Evidently when this stalagmite was formed the in-
flux of water-borne materials had ceased, or nearly so; and
Mr. Pengelly appears to affirm, though without assigning any

---

* Mr. Pengelly declines to admit this; but assigns no cause for the
breaking up of portions of the old floor, which he merely refers in general
terms to "natural causes."

reason, that none of these bones could, like the masses of stalagmite, have been lifted from lower beds, or washed into the cave from without.

The next bed marks a new change. It is a layer of black mould from three to ten inches thick. Its microscopic structure does not seem to have been examined; but it is probably a forest soil, introduced by growth, by water, by wind, and by ingress of animals, all of them modern, and contains works of art from the old British times before the Roman invasion up to the porter bottles and dropped halfpence of modern visitors. Lastly, in and upon the black mould are many fallen blocks from the roof of the cave.

There can be no doubt that this cave and the neighboring one of Brixham have done very much to impress the minds of British geologists with ideas of the great antiquity of man ; and they have, more than any other post-glacial monuments, shown the existence of some animals now extinct up to the human age. Of precise data for determining time, they have, however, given nothing. ˙ The only measures which seem to have been applied, namely, the rate of growth of stalagmite and the rate of erosion of neighboring valleys, are, from the very sequence of the deposits, obviously worthless ; and the only apparently constant measure, namely, the fall of blocks from the roof, seems not to have been applied, and Mr. Pengelly declares that it can not be practically used. We are therefore quite uncertain as to the number of centuries involved in the filling of this cave, and must remain so until some surer system of calculation can be devised. We may, however, attempt to sketch the series of events which it indicates.

The animals found in Kent's Hole are all "post-glacial," some of them of course survivors from "pre-glacial" times, and some of them still surviving. They therefore inhabited the country after it rose from the great glacial submergence. Perhaps the first colonists of the coast of Devonshire in this period were the cave bears, migrating on floating ice, and subsisting like the arctic bear and the black bear of Anti-costi, on fish, and on the garbage cast up by the sea. They may have found Kent's Hole a sea-side cavern, with perhaps some of its galleries still full of water and filling with breccia, with which the bones of dead bears became mixed. In the case of such a deposit as this breccia, however, the precise time when its materials were finally laid down in their present form, or the length of time necessary for its accumulation, can not be definitely settled. It may be a result of continued torrential action or of some sudden cataclysm. As the land rose, these creatures for the most part betook themselves to lower levels, and in process of time the cavern stood upon a hill-side, perhaps several hundreds of feet above the sea; and the mountain streams, their beds not yet emptied of glacial detritus, washed into it stones and mud, and probably bones also, while it appears that hyenas occupied the cave at intervals, and dragged in remains of mammals of many species which had now swarmed across the plains elevated out of the sea, and multiplied in the land. This was the time of the cave earth; and before its deposit was completed, though how long before an unstratified and therefore probably often-disturbed bed of this kind can not tell, man himself seems to have been added to the inhabitants of the British land. In pursuit of game he sometimes ascended the valleys beyond the cavern, or even penetrated into its outer chambers; or perhaps there were even in those days rude and savage hill-men, inhabiting the forests and warring with the more cultivated denizens of plains below, which

are now deep under the waters. Their weapons, and other implements dropped in the cavern or lost in hunting, or buried
in the flesh of wounded animals which crept to the streams
to assuage their thirst, are those found in the cave earth.
The absence of the human bones may merely show that the
mighty hunters of those days were too hardy, athletic, and
intelligent often to perish from accidental causes, and that
they did not use this cavern for a place of burial. The fragments of charcoal show that they were acquainted with fire,
and possibly that they sometimes took shelter in the cave.
But the land again subsided. The valley of that now name-
less river, of which the Rhine and the Thames may have
alike been tributaries, disappeared under the sea ; and perhaps some tribe, driven from the lower lands, took up its
abode in this cave, now again near the encroaching waves,
and left there the remains of their last repasts ere they were
driven farther inland or engulfed in the waters. For a time
the cavern may have been wholly submerged, and the charcoal of the extinguished fires became covered with its thin
coating of clay. But ere long it re-emerged to form part of
an island, long barren and desolate ; and the valleys having
been cut deeper by the receding waters, it no longer received
muddy deposits, and the crust formed by drippings from its
roof contained only bones and pebbles washed by rains and
occasional land floods from its own clay deposits. Finally,
the modern forests overspread the land, and were tenanted
by the modern animals. Man returned to use the cavern
again as a place of refuge or habitation, and to leave there
the relics contained in the black earth. This seems at present the only intelligible history of this curious cave and others resembling it ; though, when we consider the imperfection
of the results obtained even by a large amount of labor, and

the difficult and confused character of the deposits in this and similar caves, too much value should not be attached to such histories, which may at any time be contradicted or modified by new facts or different explanations of those already known. The time involved depends very much on the answer to the question whether we should regard the post-glacial subsidence and re-elevation as somewhat sudden, or as occupying long ages at the slow rate at which some parts of our continents are now rising or sinking.

Mr. Pengelly thinks it possible, but not proved, that the lower breccia of Kent's Cavern may be interglacial or preglacial in age. One case only is known where a human bone has been found in a cavern under deposits supposed to be of the nature of the glacial drift. It is that of the Victoria Cave, at Settle, in Yorkshire. At this place a human fibula was found under a layer of boulder clay. But there are too many chances of this bone having come into this position by some purely local accident to allow us to attach much importance to it until future discoveries shall have supplied other instances of the kind.*

I may close this survey of the cave deposits with a summary of the results of M. Dupont, as obtained from two of the caves explored by him, that of Margite and that of Frontal. In the first of these caverns, resting on rolled pebbles which covered the floor, were four distinct layers of river mud deposited by inundations, and amounting to two yards and a half in thickness. In all of these layers were bones. The lowest contained rude flint implements, and bones of the mammoth, rhinoceros, bear, horse, chamois, reindeer,

---

* This whole subject of supposed preglacial or interglacial men is still in great confusion and uncertainty, and is complicated with questions, still debated, as to the ages of the supposed glacial and post-glacial deposits.

stag, and hyena. In the overlying deposits are some flint implements of more artistic form and a greater prevalence of the bones of the reindeer. In the second cave, that of Frontal, over a similar deposit of alluvial mud of the mammoth age, was found a sepulchre containing the remains of sixteen individuals, of the second or diminutive Lappish race before referred to. The door of the cave had been closed by these people with a slab of stone, and in front was a hearth for funeral feasts, built on the deposits of the mammoth age, and containing bones of animals all recent or now living in Belgium, and without any traces of the bones of the extinct quadrupeds. This burial-place belonged to the Neocosmic yet prehistoric race which replaced the Palæocosmic men of the mammoth age.

What is the absolute antiquity of the Palæocosmic age in Europe? We have no monumental or historical chronology to answer this question, but only the measures of time furnished by the accumulation of deposits, by the deposition of stalagmite, by the gradual extinction of animals, and by the erosion of valleys and other physical changes. These somewhat loose measures have been applied in various ways, but the tendency of geologists, from the prevalence of uniformitarian views, and the prejudice created by familiarity with the long times of previous geologic periods, has been to assign to them too great rather than too little value, both as measures of time and as indicating a remote antiquity.

With reference to the accumulation of deposits, whether derived from disintegration of the roof and walls of the cave, introduced by land floods or river inundations or by the residence of man, their rate is of very difficult estimation. Loose stones fallen from the roof, as in the case of Kent's Cave, would give a fair measure of time if we could be sure that the climate had continued uniform, and that there had been no violent earthquakes. Mr. Pengelly has, however, hopelessly given up this kind of evidence. Where, as in the case of many of these caves, land floods and river inundations have entered, these may have been frequent or separated by long intervals of time, and they may have been of great or small amount. Where, for instance, as in one of the Belgian caves, there are six beds of ossiferous mud, but for the fact that five layers of stalagmite separate them we might not have known whether they represent six annual inundations, or floods separated by many centuries from each other.

In the case of the Victoria Cave at Settle, Dawkins, reasoning from the accumulation of two feet of detritus over British remains that may be supposed to be 1200 years old, gives a basis which would at the same rate of deposit allow about 5000 years for the date of palæolithic men; but Prestwich and others, on the basis of stalagmite deposits, claim a vastly higher antiquity for the men who made the implements found in Kent's Hole and Brixham.

If we now turn to these stalagmite floors, when we consider that they have been formed by the slow solution of limestone by rain-water charged with carbonic acid, and the dropping of this water on the floor, and when we are told that in Kent's Cavern a marked date shows that the stalagmite has grown at the rate of only one twentieth of an inch since 1688, and that there are two beds of stalagmite, one of which is in some places twelve feet thick, we are impressed with the conviction of a vast antiquity. But when we are told by Dawkins that the rate of deposit in Ingleborough Cave may be estimated at a quarter of an inch per annum, and when we consider that the present rate of deposit in Kent's Hole is probably very different from what it was in the former condition of the country, stalagmite becomes a very unsafe measure of time. With respect again to the accumulation of kitchen-midden stuff in the course of the occupancy of caverns, this proceeds with great rapidity, when caves are steadily occupied and it is not the practice to cleanse out the débris of fires, food, and bedding. Even when the occupation is temporary, a tribe of savages engaged with the preparation of dried meat and pemmican in a very short time produce a considerable heap of bones and other rejectamenta.

Looking next to the extinction of animals, we find that the species found in the oldest deposits containing human remains are in part still extant. Others which are locally extinct we know existed in Europe until historical times, that is, within the last two thousand years. How long previously to this the others became extinct we have no certain means of knowing, though it seems probable that they disappeared gradually and successively. We have, however, farther to bear in mind the possibility of cataclysms or climatal changes which may have proved speedily fatal to many species over large areas. In any case we have this certain fact that, though the time elapsed has been sufficient for the extinction of many species, it does not seem to have sufficed to effect any noteworthy change on those that survived. Farther, we may consider that time is only one factor in this matter, and not the one which is the efficient cause of change, since we know no reason why one species of animal should not continue to be reproduced as long as another, but for the occurrence of physical changes of a prejudicial character.

We have still remaining the changes which have taken place in the erosion of valleys since the caverns were occupied. Dupont informs us that the openings of some of the caverns once flooded by rivers are now in limestone cliffs two

hundred feet above the water, while no appreciable lowering
of the bottoms of the ravines is taking place now. This would
in some contingencies put back the period of filling of the
caves to an indefinite antiquity. But then the questions
occur—Was there once more water in the rivers or more ob-
struction at their outlets, or was the erosive power greater at
one time than now, or were the river valleys excavated in still
more ancient time, and partly filled with mud when the water
entered the caves, and may this mud have been since swept
away? So, in like manner, the waters flowing in the channels
near Brixham Cave and Kent's Hole were apparently about
seventy feet higher in times of flood than at present, but the
time involved is subject to the same doubts as in the case of
the Belgian caves. Hughes has well remarked that elevations
of the land, by causing rivers to form waterfalls and cascades,
which they cut back, may greatly accelerate the rate of erosion. Farther, there is the best reason to believe that in the
glacial period many old valleys were filled with clay, and that
the modern cutting consisted merely in the removal of this
clay. Belt has shown in a recent paper* good reason to believe that this is the case with the Falls of Niagara, and that
the cutting actually effected through rock within the later Pleistocene and modern period has been that only of the new
gorge from the whirlpool to Queenstown, the main part of the
ravine being of older date and merely re-excavated. This
would greatly reduce the ordinary estimate of time based on
the cutting of the Niagara gorge.

This leads us next to consider the occurrence of human remains and objects of art in the river-gravels themselves, and
the amount of excavation and deposit involved in the deposi-

* *Quarterly Journal of Science*, April, 1875.

tion of these gravels. In the river-gravels of the Somme, and of many other rivers in France and Southern England, chipped flints and rude flint implements are found in so great quantity as to imply that the beds and banks of these streams were resorted to for flint material, and that the unfinished and rejected implements left in the holes and trenches, or on the heaps where the work was carried on, were afterward sorted by running water, perhaps in abnormal floods and debacles, such as occur in all river valleys occasionally, perhaps in that great diluvial catastrophe which seems to have terminated the residence of Palæocosmic man in Europe. Wilson has well shown how the heaps left by American tribes in and near their flint quarries would furnish the material for such accumulations. The time required for the erosion of the valleys and the deposit of the gravels has been very variously estimated. In the case of the Somme, which river is not appreciably deepening its bed, if we suppose it to have cut its wide valley to the depth of one hundred and fifty feet out of solid chalk since the so-called " high level" gravels of France and the South of England were deposited, the time required shades off into infinity. So Evans, in his work on " The Ancient Stone Implements of Great Britain," looking upon the amount of excavation of wide and deep valleys since the stone implements of Bournemouth are supposed to have been deposited in gravel, says, " Who can fully comprehend how immensely remote was the epoch when that vast bay was high and dry land?" and he becomes poetical in delineating the view that must have met the eyes of " palæolithic" man. And undoubtedly, if one is to be limited to the precise nature and amount of causes now at work in the district, the time must not only be " immensely remote," but illimitably so. The difficulty lies with the exaggerated uniformitarianism of the supposition that

O

such causes could have produced the results. But, for reasons to be immediately stated, the time required is liable to numerous deductions; and recently Tylor, Pattison, Collard, and others have insisted ably on these deductions, as has also Professor Hughes, of Cambridge. I have myself urged them strongly in the work already referred to.

In the first place, when we see a deep river valley in which the present stream is doing an almost infinitesimal amount of deepening, we are not to infer that this represents all its work past and present. In times of unusual flood it may do in one week more than in many previous years. Farther, if there have been elevations or depressions of the land, when the land has been raised the cutting power has at once been enormously increased, and when depressed it has been diminished, or filling has taken the place of cutting. Again, if the climate in time past has been more extreme, or the amount of rainfall greater, the cutting action has then been proportionally rapid. Perhaps no influence is greater in this respect than that which is known to the colonists in Northeastern America as "ice-freshets," when in spring, before the ice has had time to disappear from the rivers, sudden thaws and rains produce great floods, which rushing down over the icy crust, or breaking and hurling its masses before them, work terrible havoc on the banks and alluvial flats, depositing great beds of gravel, and sweeping away immense masses that had lain undisturbed for centuries. Now we know that in Europe the human period was preceded by what has been termed the glacial age, and as it was passing away there must have been unexampled floods and ice-freshets, and a temporary "pluvial period," as it has been called, in which the volume of the rivers was immensely increased. Farther, it is an established fact that the period of the appearance of man was a time when the continents in the northern hemisphere were more elevated than at present, and when consequently the cutting action of rivers was at a maximum. This was again followed by a period of depression, accompanied probably by many local cataclysms, if not by a general deluge; and there are strong geological reasons to believe that this convulsion was connected with the disappearance from Europe of Palæocosmic man, and many of the animals his contemporaries. This view I advocated some time ago in my "Story of the Earth;" and more recently Mr. Pattison, in an able paper read before the Victoria Institute, has developed it in greater detail, and supported it by a great mass of geological authority. If the Palæocosmic period was one of continental elevation, when the greater seats of population were in the valleys of great rivers now covered by the German Ocean and the English Channel, and when the valleys of the Thames and the Somme were those of upland streams frequented by straggling parties and small tribes, and the seats of extensive flint factories for the supply of the plains below, and if this state of things was terminated by a diluvial debacle, we can account for all the phenomena of the drift implements without any extravagant estimate of time.

I quote with much pleasure on this subject the following from the report of a lecture on "Geological Measures of Time," by Professor Hughes, before the Royal Institution of London. Hughes was, like myself, a companion of Sir Charles Lyell in some of his journeys, though belonging to a younger generation of geologists, and is an accurate observer and reasoner.

"Another method of estimating the lapse of time is founded upon the supposed rate at which rivers scoop out their channels. Although no very exact estimates have been attempted, still the immense quantity of work that has been done, as compared with the slow rate at which a river is now excavating that same part of the valley, is often appealed to as a proof of a great lapse of time.

"The fact of such an enormous lapse of time is not questioned, but this part of the evidence is challenged.

"The previous considerations of the rate of accumulation of silt on the low lands prepares us to inquire whether there is any waste at all along the alluvial plains. Several examples were given to show that the lowering of valleys was brought about by receding rapids and waterfalls; for instance, following up the Rhine, its terraces could often be traced back to where the waterfall was seen to produce at once almost all the difference of level between the river reaches above and below it. At Schaffhausen the river terrace below the hotel could be traced back and found to be continuous with the river margin above the fall. The wide plains occurring here and there, such as the Mayence basin, were due to the river being arrested by the hard rocks of the gorges below Bingen so long that it had time to wind from side to side through the soft rocks above the gorges. When waterfalls cut back to such basins or to lakes they would recede rapidly, tapping the waters of the lake, eating back the soft beds of the alluvial plains, and probably in both cases leaving terraces as evidence, not of upheavals or of convulsions, but of the arrival of a waterfall which had been gradually travelling up the valley. So when the Rhone cuts back from the falls at Belgarde we shall have terraces where now is the shore of Geneva; so also when the Falls of Schaffhausen, and ages afterward when the Falls of Laufenburg have tapped the Lake of Constance, there will be terraces marking its previous levels. And so we may explain the former greater extent of the Lake of Zurich, which stood higher and spread wider by Utznach and Wetzikon before it was tapped by the arrival of waterfalls, which cut back into it and let its waters run off until they fell to their present level.

" A small upheaval near the mouth of a river would have a similar effect. The Thames below London and the Somme below St. Acheul can now only just hand on the mud brought down from higher ground; but suppose an elevation of a hundred feet over those parts of England and France (quite imperceptible if extended over 10,000, 1000, or even 100 years), and the rivers would tumble over soft mud and clay and chalk, and soon eat their way back from Sheppey to London, and from St. Valery to Amiens.

" So when we want to estimate the age of the gravels on the top of the cliff at the Reculvers, or on the edge of the plateau of St. Acheul, we have to ask, not how long would it take the rivers to cut down to their present level from the height of those gravels at the rate at which that part of their channel is being lowered now, but how long would it take the Somme or Thames, which once ran at the level of those gravels, to cut back from where its mouth or next waterfall was then to where it runs over rapids now. We ought to know what movements of upheaval and depression there have been ; what long alluvial flats or lakes which may have checked floods, but also arrested the rock protecting gravel ; how much the wash of the estuarine waves has helped. In fact, it is clear that observations made on the action of the rivers at those points now have nothing to do with the calculation of the age of the terraces above, and that the circumstances upon which the rate of recession of the waterfalls and rapids depends are so numerous and changeable that it is at present unsafe to attempt any estimate of the time required to produce the results observed."

I may close this discussion by quoting from the paper of my friend Mr. Pattison, already referred to, the following summing up of his conclusions, in which I fully concur :

"We may assume it as established that there was a time when England was connected with the Continent, when big animals roamed in summer up the watercourses and across the uplands, and man, armed only with rude stones, followed them into the marshes and woods, hunted them for sustenance, and consumed them in shelter of caves, then accessible from the river levels. This state of things was continued until disturbed by oscillations of surface, accompanied by excessive rainfalls and rushes of water from the water-sheds of the rivers, until the great animals were driven out or destroyed, and man ceased to visit these parts. The disturbances continued, the Strait of Dover was formed, the configuration of the soft parts of the islands and continents was fixed, action subsided, and the present state of things obtained. Man resumed his residence, but with loss of the mammoth and its companions. The reindeer now constituted the type of a state of things which lasted down to the historic period, without any other from that time to this. * * *

"Chronologists are agreed that about 2000 years B.C. Abraham migrated from Mesopotamia to Canaan, and that at this time Egypt at least was old in civilization. Beyond this we have no positive scale of time in Scripture ; for it is evident, from the narrative itself, that the latter does not cover the whole time. * * *

"Ussher estimates from Scripture the creation of man as about 2000 years before this. During the latter portion of this time civilization was proceeding under settled governments in the East, interrupted, says the record and tradition, by a flood. * * *

" So Lucretius :

> ' Thus, too, the insurgent waters once o'erpowered,
> As fables tell, and deluged many a state ;
> Till, in its turn, the congregated waves
> By cause more potent conquered, heaven restrain'd
> Its ceaseless torrents, and the flood decreased.'

Barbarism covered the whole Western world ; neither in the 2000 years before Abraham, nor in the 2000 years afterward, have we any light reflected from these regions to the East. In this 4000 years, or in the somewhat longer period which probably will be ultimately settled as warranted by the record, we place hypothetically all the phenomena of the later mammalian age, including the introduction of man as a hunter, the first occupation of the caves by him also, the diluvial phenomena of the wide valleys, the oscillations and disturbances of the earth's crust, alterations in the coast-line, and physical settlement of the country ; after this comes the second occupation of the caves. In short, if we say that, hypothetically, the whole first known human age occurred within 4000 years of the Christian era, no one can say that it is geologically impossible. Who can say that 1643 years is insufficient to comprise all the phenomena that occurred during a period confessedly characterized by more rapid and extensive action than at present—a period during which ruptures in the earth's crust, oscillations, and permanent uprising took place, and the intermittent action of violent floods caused the deposit and disturbance and resettlement of the gravels and brick-earth ? There is nothing to interfere with the prevalent opinion that man was introduced here while the glacial period was dying out, and while it was still furnishing floodwaters sufficient to scour and re-sort the gravels of the valleys down which they flowed. This supposition may be extended to both the great continents."

To conclude : Our mode of reconciling the Mosaic history of antediluvian man with the disclosures of the gravels and caves would be to identify Palæocosmic man, or man of the mammoth age, with antediluvian man; to suppose that the changes which closed his existence in Europe as well as Western Asia were those recorded in the Noachian deluge; and that the second colonization of the diminished and shrunken Europe of the modern period was effected by the descendants of Noah.   It may be asked—Must we suppose that the Adam of the Bible was of the type of the coarsely featured and gigantic men of the European caverns? I would answer—Not precisely so; but it is quite possible that Adam may have been Turanian in feature. We should certainly suppose him to have been a man well developed in brain and muscle.   Such men as those found in the caves would rather represent the ruder " Nephelim," the "giants that were in those days," than Adam in Eden.   Farther, the new colonists of Europe after the deluge would no doubt be a very rude and somewhat degenerate branch of Noachidæ, probably driven before more powerful tribes in the course of the dispersion.   The higher races of both periods are probably to be looked for in Western Asia; but even there we must expect to find cave men like those whose remains were found by Tristram in the caves near Tyre, and like the Horim of Moses; and we must also expect to find the antediluvian age in the main an age of stone everywhere, and its arts, except in certain great centres of population, perhaps not more advanced than those of the Polynesians, or those of the agricultural American tribes before the discovery of America by Columbus.

As a geologist, and as one who has been in the main of the school of Lyell, and after having observed with much care the deposits of the more modern periods on both sides of the Atlantic, I have from the first dissented from those of my scientific brethren who have unhesitatingly given their adhesion to the long periods claimed for human history,· and have maintained that their hasty conclusions on this subject must bring geological reasoning into disrepute, and react injuriously on our noble science. We require to make great demands on time for the pre-human periods of the earth's history, but not more than sacred history is willing to allow for the modern or human age.

O 2

CHAPTER XV.

COMPARISONS AND CONCLUSIONS.

"Lo, these are but the outlines of his ways, and how faint the whisper which we hear of him—the thunder of his power who could understand?" —Job xxvi., 14.

IN the preceding pages I have, as far as possible, avoided that mode of treating my subject which was wont to be expressed as the "reconciliation" of Scripture and Natural Science, and have followed the direct guidance of the Mosaic record, only turning aside where some apt illustration or coincidence could be perceived. In the present chapter I propose to inquire what the science of the earth teaches on these same subjects, and to point out certain manifest and remarkable correspondences between these teachings and those of revelation. Here I know that I enter on dangerous ground, and that if I have been so fortunate as to carry the intelligent reader with me thus far, I may chance to lose him now. The Hebrew Scriptures are common property; no one can fairly deny me the right to study them, even though I do so in no clerical or theological capacity; and even if I should appear extreme in some of my views, or venture to be almost as enthusiastic as the commentators of Homer, Shakespeare, or Dante, I can not be very severely blamed. But the direct comparison of these ancient records with results of modern science is obnoxious to many minds on different grounds; and all the more so that so few men are at once students both of nature

and revelation. There are, as yet, but few even of educated ⸌
men whose range of study has included any thing that is prac- ⁄
tical or useful either in Hebrew literature or geological science. That slipshod Christianity which contents itself with
supposing that conclusions which are false in nature may be
true in theology is mere superstition or professional priestcraft, and has nothing in common with the Bible; but there
are still multitudes of good men, trained in the verbal and
abstract learning which at one time constituted nearly the
whole of education, who regard geology as a mass of crude
hypotheses destitute of coherence, a perpetual battle-ground
of conflicting opinions, all destined in time to be swept away.
It must be admitted, too, that from the nature of geological
evidence, and from the liability to error in details, the solidity
of its conclusions is not likely soon to be appreciated as fully
as is desirable by the common mind; while it is unfortunately
true that the outskirts of science are infested with hosts of
half-informed and superficial writers, who state these conclusions incorrectly, or apply them in an unreasonable manner
to matters on which they have no bearing. On the other
hand, the geologist, fully aware of the substantial nature of
the foundations of the science of the earth, regards it as little
less than absurd to find parallels to its principles in an ancient theological work. Still there are possible meeting-points
of things so dissimilar as Bible lore and geological exploration.
If man is a being connected on the one hand with material
nature, and on the other with the spiritual essence of the
Creator; if that Creator has given to man powers of exploring
and comprehending his plans in the universe, and at the same
time has condescended to reveal to him directly his will on
certain points, there is nothing unphilosophical or improbable
in the supposition that the same truths may be struck out on

the one hand by the action of the human mind on nature, and on the other by the action of the Divine mind on that of man. The highest and most nobly constituted minds have ever been striving to scale heaven above and dive into the earth below, that they may extort from them the secret of their origin, and may find what are the privileges and destinies of man himself. They have learned much; and if through other gifted minds, and through his heaven-descended Word and Spirit, God has condescended to reveal himself, there must surely be much in common in that which God's works teach to earnest inquirers and that which he directly makes known. But few of our greatest thinkers, whether on nature or theology, have reached the firm ground of this higher probability; or if they have reached it, have dreaded the scorn of the half-learned too much to utter their convictions. Still this is a position which the enlightened Christian and student of nature must be prepared to occupy, humbly and with admission of much ignorance and incapacity, but with bold assertion of the truth that there are meeting-points of nature and revelation which afford legitimate subjects of study.

In entering on these subjects, we may receive certain great truths·in reference to the history of the earth as established by geological evidence. In the present rapidly progressive state of the science, however, it is by no means easy to separate its assured and settled results from those that have been founded on too hasty generalization, or are yet immature; and at the same time to avoid overlooking new and important truths, sufficiently established, yet not known in all their dimensions. In the following summary I shall endeavor to present to the reader only well‑ascertained general truths, without indulging in those deviations from accuracy for effect too often met with in popular books. On the other hand, we have already found that the Scriptures enunciate distinct doctrines on many points relating to the earth's early history, to which it will here be necessary merely to refer in general terms. Let us in the first place shortly consider the conclusions of geology as to the origin and progress of creation.

1. The widest and most important generalization of modern geology is that all the materials of the earth's crust, to the greatest depth that man can reach, either by actual excavation or inference from superficial arrangements, are of such a nature as to prove that they are not, in their present state, original portions of the earth's structure; but that they are the results of the operation, during long periods, of the causes of change—whether mechanical, chemical, or vital—now in operation, on the land, in the seas, and in the interior of the earth. For example, the most common rocks of our continents are conglomerates, sandstones, shales, and slates; all of which are made up of the débris of older rocks broken down into gravel, sand, or mud, and then re-cemented. To these we may add limestones, which have been made up by the accumulation of corals and shells, or by deposits from calcareous springs; coal, composed of vegetable matter; and granite, syenite, greenstone, and trap, which are molten rocks formed in the manner of modern lavas. So general has been this sorting, altering, and disturbance of the substance of the earth's crust, that, though we know its structure over large portions of our continents to the depth of several miles, the geologist can point to no instance of a truly primitive rock which can be affirmed to have remained unchanged and *in situ* since the beginning.

" All are aware that the solid parts of the earth consist of distinct substances, such as clay, chalk, sand, limestone, coal, slate, granite, and the like; but, previously to observation, it is commonly imagined that all had remained from the first in the state in which we now see them—that they were created in their present forms and in their present position. The geologist now comes to a different conclusion; discovering proofs that the external parts of the earth were not all produced in the beginning of things in the state in which we now behold them, nor in an instant of time. On the contrary, he can show that they have acquired their actual condition and configuration gradually and at successive periods, during each of which distinct races of living beings have flourished on the land and in the waters; the remains of these creatures lying buried in the crust of the earth."*

2. Having ascertained that the rocks of the earth have thus been produced by secondary causes, we next affirm, on the evidence of geology, that a distinct order of succession of these deposits can be ascertained; and though there are innumerable local variations in the nature of the rocks formed at the same period, yet there is, on the great scale, a regular sequence of formations over the whole earth. This succession is of the greatest importance in the case of aqueous rocks, or those formed in water; and it is evident that in the case of beds of sand, clay, etc., deposited in this way, the upper must be the more recent of any two layers. This simple principle, complicated in various ways by the fractures and disturbances to which the beds have been subjected, forms the basis of the succession of "formations" in geology as deduced from stratigraphical evidence.

3. This regular series of formations would be of little value as a history of the earth were it not that nearly all the aqueous rocks contain remains of the contemporary animals and

---

* Lyell's "Manual of Elementary Geology."

plants. Ever since the earth began to be tenanted by organized beings, the various accumulations formed in the bottoms
of seas and at the mouths of rivers have entombed remains
of marine animals, more especially their harder parts, as
shells, corals, and bones, and also fragments or entire specimens of land animals and plants. Hence, in any rock of
aqueous formation, we may find fossil remains of the living
creatures that existed in the waters in which that rock was
accumulated or on the neighboring land. If in the process
of building up the continents, the same locality constituted
in succession a part of the bottom of the ocean, of an inland
sea, of an estuary, and a lake, we should find in the fossil
remains entombed in the deposits of that place evidences of
these various conditions; and thus a somewhat curious history
of local changes might be obtained. Geology affords more
extensive disclosures of this nature. It shows that as we descend into the older formations we gradually lose sight of the
existing animals and plants, and find the remains of others
not now existing; and these, in turn, themselves disappear,
and were preceded by others; so that the whole living population of the earth appears to have been several times renewed prior to the beginning of the present order of things.
This seems farther to have occurred in a slow and gradual
manner, not by successive great cataclysms or clearances
of the surface of the earth, followed by wholesale renewal.
This doctrine of geological uniformity is, however, to be understood as limited by the equally certain fact that there has
been progress and advance, both in the inorganic arrangements of the earth's surface and in its organized inhabitants,
and that there have, in geological as in historical times, been
local cataclysms and convulsions, as those of earthquakes
and volcanoes, often on a very extensive scale. Farther, there

are good reasons to believe that there have been alternations of cold or glacial periods and of warm periods, of periods of subsidence and re-elevation, and of periods of greater and less activity of certain of the leading agents of geological change. But as to the extent of these differences and their bearing on the geological history, there is still much uncertainty and difference of opinion.*

In the sediment *now* accumulating in the bottom of the waters are being buried remains of the existing animals and plants. A geological formation is being produced, and it contains the skeletons and other solid parts of a vast variety of creatures belonging to all climates, and which have lived on land as well as in fresh and salt water. Let us now suppose that by a series of changes, sudden or gradual, all the present organized beings were swept away, and that, when the earth was renewed by the power of the Creator, a new race of intelligent beings could explore those parts of the former sea basins that had been elevated into land. They would find the remains of multitudes of creatures not existing in their time ; and by the presence of these they could distinguish the deposits of the former period from those that belonged to their own. They could also compare these remains with the corresponding parts of creatures which were their own contemporaries, and could thus infer the circumstances in which they had lived, the modes of subsistence for which they had been adapted, and the changes in the distribution of land and water and other physical conditions which had occurred. This, then, is precisely the place which fossil organic remains occupy in modern geology, except that our

---

* For a full discussion of this subject, see the "Story of the Earth and Man."

present system of nature rests on the ruins, not of one previous system, but of several.

4. By the aid of the superposition of deposits and their organic remains, geology can divide the history of the earth into distinct periods. These periods are not separated by merely arbitrary boundaries, but to some extent mark important eras in the progress of our earth; though they usually pass into each other at their confines, and the nature of the evidence prevents us from ascertaining the precise length of the periods themselves, or the intervals in time which may separate the several monuments by which they are distinguished. The following table will serve to give an idea of the arrangement at present generally received, with some of the more important facts in the succession of animal and vegetable life, as connected with our present subject. It commences with the oldest periods known to geology, and gives in the animal and vegetable kingdoms the *first appearance* of each class, with a few notes of the subsequent history of the principal forms. It must, however, be borne in mind that farther discoveries may extend some classes farther back than we at present know them, and that a more detailed table, descending to orders and families, would give a more precise view of the succession of life. Farther, the several geological formations would admit of much subdivision, and are represented locally by various kinds and different thicknesses of sediment.*

The oldest fossil remains known are the Protozoa of the Laurentian rocks. In the succeeding Cambrian or Primor-

---

* Such a table, with an admirable exposition of the entire succession, as at present known, is given in the Appendix to Lyell's "Students' Manual of Geology."

TABULAR VIEW OF THE SUCCESSION OF GEOLOGICAL FORMA-
TIONS AND ORGANIC REMAINS.

PERIODS.	SYSTEMS OF FORMATIONS.	CLASSES OF ANIMALS.	PLANTS.
I. EOZOIC PERIOD.	Ancient Metamorphic rocks of Scandinavia, Canada, etc.	Eozoon and probably other Protozoa.	Graphite and Iron Ores representing Vegetable Matter.
II. PRIMARY OR PALÆOZOIC PERIOD.	Cambrian.....	*Radiata*—Hydrozoa, Echinodermata (Cystideans). *Mollusca* — Brachiopoda, Lamellibranchiata, Gasteropoda, Cephalopoda (Bivalve and Univalve Shell-fishes). *Articulata*—Annelida, Crustacea (Worms and Soft Shell-fishes of the lower grades).	Algæ.
	Lower Silurian.	*Radiata*—Anthozoa (coral animals), Echinodermata (sea stars, etc.). *Mollusca*—Polyzoa, Tunicata. Other Mollusks and Articulates as before.	Algæ.
	Upper Silurian.	Radiates, Mollusks, and Articulates as before. *Vertebrata*—First Ganoid and Placoid Fishes.	Acrogenous Land plants.
	Erian or Devonian........	*Articulata* — Insects and higher Crustaceans. *Vertebrata* — Fishes, Ganoid and Placoid.	Acrogens and Gymnosperms.

PERIODS.	SYSTEMS OF FORMATIONS.	CLASSES OF ANIMALS.	PLANTS.
II. PRIMARY OR PALÆOZOIC PERIOD *continued.*	Carboniferous..	*Mollusca* — Pulmonata (Land Snails). *Articulata* — Myriapods, Arachnidans (Gally-worms, Spiders and Scorpions). *Vertebrata* — Batrachians or Amphibians prevalent.	Acrogens, Gymnosperms, Endogens?
	Permian.......	*Vertebrata*—Lacertian or Lizard-like Reptiles.	
III. SECONDARY OR MESOZOIC PERIOD.	Triassic.......	*Vertebrata*—Higher Reptiles prevalent; Marsupial Mammals.	
	Jurassic......	*Vertebrata*—Great prevalence of higher Reptiles; Fishes, homocerque; Earliest Birds.	Endogenous trees.
	Cretaceous....	*Vertebrata*—Decadence of reign of Reptiles; Ordinary Bony fishes.	Angiospermous Exogens.
IV. TERTIARY OR CAINOZOIC PERIOD.	Eocene........	*Vertebrata* — Mammals prevalent, especially Pachyderms; Cycloid and Ctenoid Fishes prevalent. First *living* Invertebrates.	Exogens prevalent.
	Miocene.......	Living Invertebrates more numerous.	Some Modern Species appear.
	Pliocene......	Living Invertebrates still more numerous.	
V. POST-TERTIARY OR MODERN PERIOD.	Post-Pliocene..	First living Mammals. Living Invertebrates prevalent.	Existing vegetation.
	Post-Glacial and Recent......	Man and living Mammals.	

dial rocks we find many extinct species of zoophytes, shell-
fish, and crustaceans, and the algæ or sea-weeds. In the Palæozoic period as a whole, though numerous Batrachian or Am-
phibian reptiles existed toward its close, the higher orders of
fishes seem to have been the dominant tribe of animals ; and
vegetation was nearly limited to cryptogams and gymnosperms.
In the Mesozoic period, though small mammalia had been
created, large terrestrial and marine reptiles were the ruling
race, and fishes occupied a subordinate position ; while, at
the close, the higher orders of plants .took a prominent
place.   In the Tertiary and Modern eras, the mammalia,
with man, have assumed the highest or dominant position
in nature.

On this series of groups, and the succession of living beings,
Sir. C. Lyell remarks: "It is not pretended that the principal
sections called Primary, Secondary, and Tertiary are of equivalent importance, or that the subordinate groups comprise
monuments relating to equal portions of time or of the earth's
history.   But we can assert that they each relate to successive periods, during which certain animals and plants, for the
most part peculiar to their respective eras, flourished, and
during which different kinds of sediment were deposited."

We have already, in previous chapters, noticed the parallelism of the succession of life in the earth as revealed in Genesis with that disclosed by geology ; but this subject must be
farther referred to in the sequel, and in the mean time the
reader may compare for himself the succession of life in the
table with that in the later creative days.

5. The lapse of time embraced in the geological history of
the earth is enormous. Fully to appreciate this it is necessary to study the science in detail, and to explore its phenomena as disclosed in actual nature. A few facts, however, out

of hundreds which might have been selected, will suffice to indicate the state of the case. The delta and alluvial plain of the Mississippi have an area of more than 12,000 square miles, and must have an average depth of about 800 feet. At the present rate of conveyance of sediment by the river, it has been calculated that a period of about 33,000 years is implied in the deposition of this comparatively modern formation.* To be quite safe, let us take 30,000 years, and add 50,000 more for the remainder of the Post-pliocene or Quaternary. We may then safely multiply this number by forty, for the length of the Tertiary period. We may add three times as much for the Mesozoic period, and this will be far under the truth. It will then be quite safe to assume that the Palæozoic period was three times as long as the Mesozoic and Tertiary together. This would give altogether, say, 51,280,000 years for the whole of geological time from the beginning of the Palæozoic, leaving the duration of the Eozoic and previous periods undetermined, but requiring perhaps nearly as much time. Great though these demands may seem, they would be probably far below the rigid requirements of the case were it not for the probability that the present rate of transference of material by the great river is less than it was in Post-pliocene and early modern times. This might enable us to reduce our estimate considerably within the scope of a hundred millions of years.† Take another illustration from an older formation. An excellent coast section at

---

* Lyell, basing his calculations on the surveys of Messrs. Humphreys and Abbott, but others give very different estimates.

† A perfectly parallel example is that of the growth of the peninsula of Florida in the modern period, by the same processes now adding to its shores ; and this has afforded to Professor Agassiz a still more extended measure of the Post-tertiary period.

the Joggins, in Nova Scotia, exhibits in the coal formation proper a series of beds with erect trunks and roots of trees *in situ*, amounting to nearly 100. About 100 forests have successively grown, partially decayed, and been entombed in muddy and sandy sediment. In the same section, including in all about 14,000 feet of beds, there are 76 seams of coal, each of which can be proved to have taken more time for its accumulation than that required for the growth of a forest. Supposing all these separate fossil soils and coals to have been formed with the greatest possible rapidity, forty thousand years would be a very moderate calculation for this portion of the Carboniferous system; and for aught that we know thousands of years may be represented by a single fossil soil. But this is the age of only one member of the Carboniferous system, itself only a member of the great Palæozoic group, and we have made no allowance for the abrasion from previous rocks and deposition of the immense mass of sandy and muddy sediment in which the coals and forests are imbedded, and which is vastly greater than the deltas of the largest modern rivers.

Considerations of a physical rather than of a geological nature also give us long periods for the probable existence of the earth, though they serve to correct somewhat the extravagant estimates of some theorists. Croll has based an interesting calculation on the amount of erosion of the land by rivers. That of the Mississippi amounts to one foot in 6000 years. That of the Ganges gives one foot in 2358 years, the average being, say, one foot in 4179 years. Some smaller rivers give a much shorter time; but the average of two great rivers, one draining a very large area of the western and another of the eastern hemisphere, and in very different climates and geographical conditions, will probably be the most reliable datum.

Croll, however, prefers the Mississippi rate.* If we estimate the proportion of land to water as 576 to 1390, this will give for the entire area of the ocean a rate of deposition of one foot in 14,400 years. Now the entire thickness of all the stratified rocks is estimated at 72,000 feet; and at this rate the enormous time of 1,036,800,000 years would be necessary. But we have no right to assume that deposition has been going on uniformly over the entire sea-bottom. On the contrary, the greater part of it takes place within a belt of about one hundred miles from the coasts, and the deposit of calcareous and other matters over the remainder will scarcely make up for the portions of this belt on which no deposit is taking place. This will give an area of deposit of about 11,650,000 square miles, consequently only one twelfth of the above time, or about 86,400,000 years, would be required. This can be but a very rough calculation; but it has the merit of squaring very nearly with the calculations derived from physical considerations, more especially by Sir William Thomson, which limit the possible existence of the earth's solid crust to one hundred millions of years. Similar conclusions have also been deduced from what is known of the physical constitution of the sun. Croll's own ingenious theory of glacial periods produced by the varying eccentricity of the earth's orbit, along with the precession of the equinoxes, would give, according to him, about 80,000 years ago for the date of the Glacial period, and for the beginning of the Tertiary period about 3,000,000 years ago.

It would thus appear that physical and geological science

* Reade, of Liverpool, has recently given a much slower rate—one foot in 13,000 years—as a result of recent English surveys; but I have not seen his precise data, and the result certainly differs from those of all other observations.

conspire in assigning a great antiquity to the earth, but not
an unlimited antiquity. They agree in restricting the ages
that have elapsed since the introduction of life within one
hundred millions of years. I confess, however, that a consideration of the fact that all our geological measures of erosion and deposition seem to be based on cases which refer
to what may be termed minimum action leads me to believe
that the actual time will fall very far within this limit. For
example, if we were to suppose an elevation of the land
drained by the Mississippi even to a small amount, its cutting
power would be vastly increased for a long time. The same
effect would result from a subsidence and re-elevation, or
from any cause increasing the amount of rainfall or deposition of snows in winter. Now we know that such things
have occurred in the past, while we have no reason to believe that the amount of action was ever much less than at
present. Similar considerations apply to nearly all our geological measures of time ; and there has been a tendency
to exaggerate these, as if geologists were entitled to demand
unlimited time, and to stretch the doctrine of uniformity to
the utmost.

6. During the whole time referred to by geology, the great
laws both of inorganic and organic nature have been the same
as at present. The evidence of light and darkness, of sunshine and shower, of summer and winter, and of all the known
igneous and aqueous causes of change, extends back almost,
and in some of these cases altogether, to the beginning of the
Palæozoic period. In like manner the animals and plants
of the oldest rocks are constructed on the same physiological
and anatomical principles with existing tribes, and they can
be arranged in the same genera, orders, or classes, though
specifically distinct. The revolutions of the globe have in-

volved no change of the general laws of matter ; and though it is possible that geology has carried us back to the time when the laws that regulate life began to operate, it does not show that they were less perfect than now, and it indicates no trace of the beginning of the inorganic laws. Geological changes have resulted not from the institution of new laws, but from new *dispositions*, under existing laws and general arrangements. There is every reason to believe that in the inorganic world these dispositions have required no new creative interpositions during the time to which geology refers, but merely the continued action of the properties bestowed on matter when first produced. In the organic world the case is different.

7. In the succession of animal and vegetable life we find a constant improvement and advance by the introduction of new types of being. We have already given a general outline of this advancement of organized nature. It has consisted in the introduction, from time to time, of new and more highly organized beings, so as at once to increase the variety of nature, and to provide for the elevation of the summit of the graduated scale of life to higher and higher points. At the same time, in each successive period, it has been the law of creation that the forms of life then dominant should attain their highest development, and should then be succeeded by more advanced types. For instance, in the earlier Palæozoic period we have molluscous animals and fishes, then apparently the highest forms of life, appearing with a very advanced organization, not surpassed, if even equalled, in modern times. In the latter part of the same period, some lower forms of vegetable life, now restricted to a comparatively humble place, were employed to constitute magnificent forests. In the Mesozoic period, again, reptiles attained to their highest point

P

338 *The Origin of the World.*

in organization and variety of form and employment, while mammalia had as yet scarcely appeared.*

8. If now we ask in what manner the succession of life on the earth has been produced, two apparently opposite hypotheses rise before us. The one is that of introduction of new species by creative acts, the other that of development of new species by changes of those previously existing. In one respect the difference of these views is little more than one of expression, for the meaning of the statements depends on what we understand by a species and what by a mere varietal form, and also on what we understand by creation and what we mean by development. Twenty years ago nearly all geologists were believers in creation, though it must be admitted without precisely understanding what they meant by the term. Now, the great impression produced by Darwin's speculations and the prevalence of the evolutionist philosophy have produced a leaning in the other direction. More recently, however, the absurdities into which the extreme evolutionists find themselves driven have produced a reaction; and we hope that views consistent with revelation, or at least with Theism, will again be in the ascendant, and that present controversies will serve to give more precise and definite views than heretofore of the relation of nature to God. As illustrations of

---

* I am quite aware that it may be objected to all this that it is based on merely negative evidence; but this is not strictly the case. There are positive indications of these truths. For example, in the Mesozoic epoch the lacertian reptiles presented huge elephantine carnivorous and herbivorous species—the Megalosaurus, Iguanodon, etc.; flying species, with hollow bones and ample wings—the Pterodactyles; and aquatic whale-like species — Pliosaurus, Ichthyosaurus, etc. These creatures actually filled the offices now occupied by the mammals; and, though lacertian in their affinities, they must have had circulatory, respiratory, and nervous systems far in advance of any modern reptiles even of the order of Loricates.

the opinions prevalent before the rise of the development theory, I may quote from Pictet and Bronn, two of the most eminent palæontologists.

Pictet says, in the introduction to his "Traité de Paléontologie:" "It seems to me impossible that we should admit, as an explanation of the phenomena of successive faunas, the passage of species into one another; the limits of such transitions of species, even supposing that the lapse of a vast period of time may have given them a character of reality much greater than that which the study of existing nature leads us to suppose, are still infinitely within those differences which distinguish two successive faunas. Lastly, we can least of all account by this theory for the appearance of new *types*, to explain the introduction of which we must necessarily, in the present state of science, recur to the idea of distinct creations posterior to the first."

The following are the general conclusions of Bronn, in his elaborate and most valuable essay, presented to the French Academy in 1856, as summarized in a notice of the work in the Journal of the Geological Society:

"1. The first productions of this power in the oldest Neptunian strata of the earth consisted of Plants, Zoophytes, Mollusks, Crustaceans, and perhaps even Fish; the simultaneous appearance of which, therefore, contradicts the assumption that the more perfect organic forms arose out of the gradual transformation in time of the more imperfect forms.

"2. The same power which produced the first organic forms has continued to operate in intensively as well as extensively increasing activity during the whole subsequent geological period, up to the final appearance of man; but here also can no traces be found of a gradual transformation of old species and genera into new; but the new have everywhere appeared as new without the co-operation of the former.

" 3. In the succession of the different forms of plants and animals, a certain regular course and plan is perceptible, which is quite independent of chance. While all species possess only a limited duration, and must sooner or later disappear, they make way for subsequent new ones, which not only almost always offer an equivalent, in number, organization, and duties to be performed, for those which have disappeared, but which are also generally more varied, and therefore more perfect, and always maintain an equilibrium with each other in their stage of organization, their mode of life, and functions. There always exists, therefore, a certain fixed relation between the newly arising and the disappearing forms of organic life.

" 4. A similar relation necessarily exists between the newly arising organic forms and the outward conditions of life which prevailed at their first appearance on the earth's surface, or at the place of their appearance.

" 5. A fixed plan appears to be the basis of the whole series of development of organic forms, in so far as man makes his first appearance at its close, when he finds every thing prepared that is necessary to his own existence and to his progressive development and improvement—which would not have been possible had he appeared at a former period.

" 6. Such a regular progress in carrying out the same plan from the beginning to the end of a period of millions of years can only be accounted for in one of two ways. Either this course of successive development during millions of years has been the regular immediate result of the systematic action of a conscious Creator, who on every occasion settled and carried out not only the order of appearance, formation, organization, and terrestrial object of each of the countless numbers of species of plants and animals, but also the number of the first individuals, the place of their settlement in every instance, although it was in his power to create every thing at once—or there existed some natural power hitherto entirely unknown to us, which by means of its own laws formed the species of plants and animals, and arranged and regulated all those countless individual conditions ; which power, however, must in this case have stood in the most immediate connection with, and in perfect subordination to, those powers which caused the gradually progressing perfection of the crust of the earth, and the gradual development of the outward conditions of life for the constantly increasing numbers and higher classes of organic forms in consequence of this perfection. Only in this way can we explain how the development of the organic world could have regularly kept pace with that of the inorganic. Such a power, although we know it not, would not only be in perfect accordance with all the other functions of nature, but the Creator, who regulated the development of organic nature by means of such a force so implanted in it, as he guides that of the inorganic world by the mere co-operation of attraction and affinity, must appear to us more exalted and imposing than if we assumed that he must always be giving the same care to the introduction and change of the vegetable and animal world on the surface of the earth as a gardener daily bestows on each individual plant in the arrangement of his garden.

" 7. We therefore believe that all species of plants and animals were originally produced by some natural power unknown to us, and not by transformation from a few original forms, and that that power was in the closest and most necessary connection with those powers and circumstances which effected the perfection of the earth's surface."

Barrande also, probably the greatest living palæontologist of Europe, adheres substantially to these views; as Agassiz did, and I believe Hall and Dana still do, in America.

I have, for my own part, seen no reason to dissent from these views, though in the sequel I shall endeavor to present some considerations which may tend to reconcile with them some of the hypotheses of a contrary nature now held. It must be admitted, however, that the majority of geologists and biologists have abandoned these views of Pictet and Bronn, and have gone over to the evolutionist philosophy, with how little reason I have endeavored to show elsewhere,* and shall farther illustrate in the Appendix. Let it be observed, however, that even evolution does not affect the grand idea of the unity of nature, or the fact that the plan of the Creator in the organic world was so vast that it required the whole duration of our planet, in all its stages of physical existence, to embrace the whole. There is but one system of organic nature; but, to exhibit the whole of it, not only all the climates and conditions now existing are required, but those also of all past geological periods. Further, the progress of nature being mainly in the direction of differentiation of functions once combined, it has a limit backward in the most general forms and conditions, and forward in the most specialized. This is the history of the individual and probably also of the type, of the world itself and of the universe; and for this reason material nature necessarily lacks the eternity of its author.

It appears, from the above facts and reasonings, that geology informs us—1. That the materials of our existing continents are of secondary origin, as distinguished from primitive

---

* "Story of the Earth"—concluding chapters.

or coeval with the beginning.   2. That a chronological order of formation of these rocks can be made out.   3. That the fossil remains contained in the rocks constitute a chronology of animal and vegetable existence.   4. That the history of the earth may be divided in this way into distinct periods, all pre-Adamite.   5. That the pre-Adamite periods were of enormous duration.   6. That during these periods the existing general laws of nature were in force, though the dispositions of inorganic nature were different in different periods, and the animals and plants of successive periods were also different from each other.   7. The introduction of new species of animals and of plants, while indicating advance in the perfection of nature, does not prove spontaneous development, but rather a definite plan and law of creation.

The parallelism of these conclusions of careful inductive inquiry into the structure of the earth's crust, with the results which we have already obtained from revelation, may be summed up under the following heads :

1. Scripture and Science both testify to the great fact that there was a beginning—a time when none of all the parts of the fabric of the universe existed ; when the Self-Existent was the sole occupant of space.   The Scriptures announce in plain terms this great truth, and thereby rise at once high above atheism, pantheism, and materialism, and lay a broad and sure foundation for a pure and spiritual theology.   Had the pen of inspiration written but the words, "In the beginning God created the heavens and the earth," and added no more, these words alone would have borne the impress of their heavenly birth, and would, if received in faith, have done much for the progress of the human mind.   These words contain a negation of hero-worship, star-worship, animal-worship, and every other form of idolatry.   They still

more emphatically deny atheism and materialism, and point upward from nature to its spiritual Creator—the One, the Triune, the Eternal, the Self-Existent, the All-Pervading, the Almighty. They call upon us, as with a voice of thunder, to bow down before that Awful Being of whom it can be said that he created the heavens and the earth. They thus embody the whole essence of natural theology, and most appropriately stand at the entrance of Holy Scripture, referring us to the works which men behold, as the visible manifestation of the attributes of the Being whose spiritual nature is unveiled in revelation. Scripture thus begins with the announcement of a great ultimate fact, to which science conducts us with but slow and timid steps. Yet science, and especially geological science, can bear witness to this great truth. The materialist, reasoning on the fancied stability of natural things, and their inscription within invariable laws, concludes that matter must be eternal. No, replies the geologist, certainly not in its present form. This is but of recent origin, and was preceded by other arrangements. Every existing species can be traced back to a time when it was not ; so can the existing continents, mountains, and seas. Under our processes of investigation the present melts away like a dream, and we are landed on the shores of past and unknown worlds. But I read, says the objector, that you can see " no evidence of a beginning, no prospect of an end." It is true, answers geology ; but, in so saying, it is not intended that the present state of things had not an ascertained beginning, but that there has been a great and, so far as we know, unlimited series of changes carried on under the guidance of intelligence. These changes we have traced back very far, without being able to say that we have reached the first. We can trace back man and his contemporaries to their origin, and we can reach the points at which still older dynasties of life began to exist. Knowing, then, that all these had a beginning, we infer that if others preceded them they also had a beginning. But, says another objector, is not the present the child of the past? Are not all the creatures that inhabit the earth the lineal descendants of creatures of past periods, or may not the whole be parts of one continual succession, under the operation of an eternal law of development? No, answers geology, species are immutable, except within narrow limits, and do not pass into each other, in tracing them toward their origin. On the contrary, they appear at once in their most perfect state, and continue unchanged till they are forced off the stage of existence to give place to other creatures. The origin of species is a mystery, and belongs to no natural law that has yet been established. Thus, then, stands the case at present. Scripture asserts a beginning and a creation. Science admits these, as far as the objects with which it is conversant extend, and the notions of eternal succession and spontaneous development, discountenanced both by theology and science, are obliged to take refuge in those misty regions where modern philosophical skepticism consorts with the shades of departed heathenism.*

2. Both records exhibit the progressive character of creation, and in much the same aspect. The Almighty might have called into existence, by one single momentary act, a world complete in all its parts. From both Scripture and geology we know that he has not done so—why we need not

---

* This was written in 1860 for the first edition of " Archaia." I see no reason to change it now, and its vindication will be found in the Appendix.

inquire, though we can see that the process employed was that best adapted to show forth the variety of his resources and the infinitely varied elements that enter into the perfect whole.

The Scripture history may be viewed as dividing the progress of the creation into two great periods, the later of which only is embraced in the geological record. ·The first commences with the original chaos, and reaches to the completion of inorganic nature on the fourth day. Had we any geological records of the first of these· periods, we should perceive the evidences of slow mutations, tending to the sorting and arrangement of the materials of the earth, and to produce distinct light and darkness, sea and land, atmosphere and cloud, out of what was originally a mixture of the whole. We should also, according to the Scriptural record, find this period interlocking with the next, by the intervention of a great vegetable creation, before the final adjustment of the earth's relations to the other bodies of our system. The second period is that of the creative development of animal life. From both records we learn that various ranks or gradations existed from the first introduction of animals; but that on the earlier stages only certain of the lower forms of animals were present ; that these soon attained their highest point, and then gradually, on each succeeding platform, the variety of nature in its higher—the vertebrate—form increased, and the upper margin of animal life attained a more and more elevated point, culminating at length in man ; while certain of the older forms were dropped, as no longer required.

In the oldest fossiliferous rocks next to the Eozoic, which so far have afforded only Protozoa—e. g., the Cambrian and Lower Silurian—we find the mollusca represented mainly by their highest and lowest classes, by allies of the cuttle-fish and nautilus, and by the lowest bivalve shell-fishes. The Articulata are represented by the highest marine class—the crustaceans—and by the lowest—the worms, which have left their marks on some of the lowest fossiliferous beds. The Radiata, in like manner, are represented by species of their highest class—the star-fishes, etc.—and by some of their simpler polyp forms. At the very beginning, then, of the fossiliferous series, the three lower sub-kingdoms exhibit species of their most elevated aquatic classes, though not of the very highest orders in those classes. The vertebrated sub kingdom has, as far as yet known, no representative in these lowest beds. In the Upper Silurian series, however, we find remains of fishes; and in the succeeding Devonian and carboniferous rocks the fishes rise to the highest structures of their class; and we find several species of reptiles, representing the next of the vertebrated classes in ascending order. Here a very remarkable fact meets us. Before the close of the Palæozoic period the three lower sub-kingdoms and the fishes had already attained the highest perfection of which their types are capable. Multitudes of new species and genera were added subsequently, but none of them rising higher in the scale of organization than those which occur in the Palæozoic rocks. Thenceforth the progressive improvement of the animal kingdom consisted in the addition, first of the reptile, which attained its highest perfection and importance in the Mesozoic period, and then of the bird and mammal, which did not attain their highest forms till the Modern period. This geological order of animal life, it is scarcely necessary to add, agrees perfectly with that sketched by Moses, in which the lower types are completed at once, and the progress is wholly in the higher.

In the inspired narrative we have already noticed some peculiarities, as, for instance, the early appearance of a highly developed flora, and the special mention of great reptiles in the work of the fifth day, which correspond with the significant fact that high types of structure appeared at the very introduction of each new group of organized beings—a fact which, more than any other in geology, shows that, in the organic department, elevation has always been a strictly *creative* work, and that there is in the constitution of animal species no innate tendency to elevation, but that on the contrary we should rather suspect a tendency to degeneracy and ultimate disappearance, requiring that the fiat of the Creator should after a time go out again to " renew the face of the earth." In the natural as in the moral world, the only law of progress is the will and the power of God. In one sense, however, progress in the organic world has been dependent on, though not caused by, progress in the inorganic. We see in geology many grounds for believing that each new tribe of animals or plants was introduced just as the earth became fitted for it ; and even in the present world we see that regions composed of the more ancient rocks, and not modified by subsequent disturbances, present few of the means of support for man and the higher animals ; while those districts in which various revolutions of the earth have accumulated fertile soils or deposited useful minerals are the chief seats of civilization and population. In like manner we know that those regions which the Bible informs us were the cradle of the human race and the seats of the oldest nations are geologically among the most recent parts of the existing continents, and were no doubt selected by the Creator partly on that account for the birthplace of man. We thus find that the Bible and the geologists are agreed not only as to the fact and order of progress, but also as to its manner and use.

3. Both records agree in affirming that since the beginning there has been but one great system of nature. We can imagine it to have been otherwise. Our existing nature might have been preceded by a state of things having no connection with it. The arrangements of the earth's surface might have been altogether different; races of creatures might have existed having no affinity with or resemblance to those of the present world, and we might have been able to trace no present beneficial consequences as flowing from these past states of our planet. Had geology made such revelations as these, the consequences in relation to natural theology and the credibility of Scripture would have been momentous. The Mosaic narrative could scarcely, in that case, have been interpreted in such a manner as to accord with geological conclusions. The questions would have arisen—Are there more creative Powers than one? If one, is He an imperfect or capricious being who changes his plans of operation? The divine authority of the Scriptures, as well as the unity and perfections of God, might thus have been involved in serious doubts. Happily for us, there is nothing of this kind in the geological history of the earth; as there is manifestly nothing of it in that which is revealed in Scripture.

In the Scripture narrative each act of creation prepares for the others, and in its consequences extends to them all. The inspired writer announces the introduction of each new part of creation, and then leaves it without any reference to the various phases which it assumed as the work advanced. In the grand general view which he takes, the land and seas first made represent those of all the following periods. So do the first plants, the first invertebrate animals, the first fishes, reptiles, birds, and mammals. He thus assures us that, however long the periods represented by days of creation, the system of nature was one from the beginning. In like manner in the geological record each of the successive conditions of the earth is related to those which precede and those which follow, as part of a series. So also a uniform plan of construction pervades organic nature, and uniform laws the inorganic world in all periods. We can thus include in one system of natural history all animals and plants, fossil as well as recent, and can resolve all inorganic changes into the operation of existing laws. The former of these facts is in its nature so remarkable as almost to warrant the belief of special design. Naturalists had arranged the existing animals and plants, without any reference to fossil species, in kingdoms, sub-kingdoms, classes, orders, families, and genera. Geological research has added a vast number of species not now existing in a living state; yet all these fossils can be inserted within the limits of recognized groups. We do not require to add a new kingdom, sub-kingdom, or class; but, on the contrary, all the fossil genera and species go into the existing divisions, in such a manner as to fill them up precisely where they are most deficient, thus occupying what would otherwise be gaps in the existing system of nature. The principal difficulty which they occasion to the zoologist and botanist is that, by filling the intervals between genera previously widely separated, they give to the whole a degree of continuity which renders it more difficult to decide where the boundaries separating the groups should be placed.

We also find that the animals and plants of the earlier periods often combined in one form powers and properties afterward separated in distinct groups; thus in the earlier formations the sauroid fishes unite peculiarities afterward divided between the fish and reptiles, constituting what Agassiz has called a synthetic type. Again, the series of creatures in time accords with the ranks which a study of their types of structure induces the naturalist to assign them in his system; and also within each of the great sub-kingdoms presents many points of accordance with the progress of the embryonic development of the individual animal. Nor is this contradictory to the statement that the earlier representatives of types are often of high and perfect organization, for the progress both in geological time and in the life of the individual is so much one of specialization that an immature animal often presents points of affinity to higher forms that disappear in the adult. In connection with this, earlier organic forms often appear to foreshadow and predict others that are to succeed them in time, as the winged and marine reptiles of the Mesozoic foreshadow the birds and cetaceans. Agassiz has admirably illustrated these links of connection between the past and the present in the essay on classification prefixed to his "Contributions to the Natural History of America." In reference to "prophetic" types, he says: "They appear now like a prophecy in those earlier times of an order of things not possible with the earlier combinations then prevailing in the animal kingdom, but exhibiting in a later period in a striking manner the antecedent consideration of every step in the gradation of animals."

4. The periods into which geology divides the history of the earth are different from those of Scripture, yet when properly understood there is a marked correspondence. Geology refers only to the fifth and sixth days of creation, or, at most, to these with parts of the fourth and seventh, and it divides this portion of the work into several eras, founded on alter-

nations of rock formations and changes in organic remains. The nature of geological evidence renders it probable that many apparently well-marked breaks in the chain may result merely from deficiency in the preserved remains; and consequently that what appear to the geologist to be very distinct periods may in reality run together.' The only natural divisions that Scripture teaches us to look for are those between the fifth and sixth days, and those which within these days mark the introduction of new animal forms, as, for instance, the great reptiles of the fifth day. We have already seen that the beginning of the fifth day can be referred almost with certainty to the Palæozoic period. The beginning of the sixth day may with nearly equal certainty be referred to that of the Tertiary era. The introduction of great reptiles and birds in the fifth day synchronizes and corresponds with the beginning of the Mesozoic period; and that of man at the close of the sixth day with the commencement of the Modern era in geology. These four great coincidences are so much more than we could have expected, in records so very different in their nature and origin, that we need not pause to search for others of a more obscure character. It may be well to introduce here a tabular view of this correspondence between the geological and Biblical periods, extending it as far as either record can carry us, and thus giving a complete general view of the origin and history of the world as deduced from revelation and science. In comparing this table with that on page 330, it will be observed that the latter refers to the last half of the creative week only, the earlier half being occupied with physical changes which, however probable inferentially, are not within the scope of geological observation.

5. In both records the ocean gives birth to the first dry land, and it is the sea that is first inhabited, yet both lead at

PARALLELISM OF THE SCRIPTURAL COSMOGONY WITH THE ASTRONOMICAL AND GEOLOGICAL HISTORY OF THE EARTH.

BIBLICAL ÆONS.	PERIODS DEDUCED FROM SCIENTIFIC CONSIDERATIONS.
The Beginning.	Creation of Matter.
First Day.—Earth mantled by the Vaporous Deep—Production of Light.	Condensation of Planetary Bodies from a nebulous mass—Hypothesis of original incandescence.
Second Day.—Earth covered by the Waters—Formation of the Atmosphere.	Primitive Universal Ocean, and establishment of Atmospheric equilibrium.
Third Day.—Emergence of Dry Land—Introduction of Vegetation.	Elevation of the land which furnished the materials of the oldest rocks—Eozoic Period of Geology?
Fourth Day.—Completion of the arrangements of the Solar System.	Metamorphism of Eozoic rocks and disturbances preceding the Cambrian epoch—Present arrangement of Seasons—Dominion of "Existing Causes" begins.
Fifth Day.—Invertebrates and Fishes, and afterward great Reptiles and Birds created.	Palæozoic Period—Reign of Invertebrates and Fishes. Mesozoic Period—Reign of Reptiles.
Sixth Day.—Introduction of Mammals—Creation of Man and Edenic Group of Animals.	Tertiary Period—Reign of Mammals. Post-Tertiary—Existing Mammals and Man.
Seventh Day.—Cessation of Work of Creation—Fall and Redemption of Man.	Period of Human History.
Eighth Day.—New Heavens and Earth to succeed the Human Epoch—"The Rest (Sabbath) that remains to the People of God." *	

* Heb. iv., 9; 2 Peter iii., 13.

Note.—The above table is identical with that published in "Archaia" in 1860, and which the author sees no reason now to change.

least to the suspicion that a state of igneous fluidity preceded
the primitive universal ocean. In Scripture the original
prevalence of the ocean is distinctly stated, and all geologists
are agreed that in the early fossiliferous periods the sea
must have prevailed much more extensively than at present.
Scripture also expressly states that the waters were the birthplace of the earliest animals, and geology has as yet discovered in the whole Silurian series no terrestrial animal, though
marine creatures are extremely abundant ; and though air-
breathing creatures are found in the later Palæozoic, they are,
with the exception of insects, of that semi-amphibious character which is proper to alluvial flats and the deltas of rivers.
It is true that the negative evidence collected by geology
does not render it altogether impossible that terrestrial animals, even mammals, may have existed in the earliest periods ; yet there are, as already pointed out, some positive
indications opposed to this. The Scripture, however, commits itself to the statement that the higher land animals did
not exist so early, though it must be observed that there is
nothing in the Mosaic narrative adverse to the existence of
birds, insects, and reptiles in the earlier Palæozoic periods.
I have said that the Bible, which informs us of a universal
ocean preceding the existence of land, also gives indications of
a still earlier period of igneous fluidity or gaseous expansion.
Geology also and astronomy have their reasonings and speculations as to the prevalence of such conditions. Here, however, both records become dim and obscure, though it is evident that both point in the same direction, and combine
those aqueous and igneous origins which in the last century
afforded so fertile ground of one-sided dispute.

6. Both records concur in maintaining what is usually termed the doctrine of existing causes in geology. . Scripture and

geology alike show that since the beginning of the fifth day, or Palæozoic period, the inorganic world has continued under the dominion of the same causes that now regulate its changes and processes. The sacred narrative gives no hint of any creative interposition in this department after the fourth day; and geology assures us that all the rocks with which it is acquainted have been produced by the same causes that are now throwing down detritus in the bottom of the waters, or bringing up volcanic products from the interior of the earth. This grand generalization, therefore, first worked out in modern times by Sir Charles Lyell, from a laborious collection of the changes occurring in the present state of the world, was, as a doctrine of divine revelation, announced more than three thousand years ago by the Hebrew lawgiver; not for scientific purposes, but as a part of the theology of the Hebrew monotheism.

7. Both records agree in assuring us that death prevailed in the world ever since animals were introduced. The punishment threatened to Adam, and considerations connected with man's state of innocence, have led to the belief that the Bible teaches that the lower animals, as well as man, were exempt from death before the fall. When, however, we find the great *tanninim*, or crocodilian reptiles, created in the fifth day, and beasts of prey on the sixth, we need entertain no doubt on the subject, in so far as Scripture is concerned. The geological record is equally explicit. Carnivorous creatures, with the most formidable powers of destruction, have left their remains in all parts of the geological series; and indeed, up to the introduction of man, the carnivorous fishes, reptiles, and quadrupeds were the lords and tyrants of the earth. There can be little doubt, however, that the introduction of man was the beginning of a change in this respect.

A creature destitute of offensive weapons, and subsisting on fruits, was to rule by the power of intellect.[+] As already hinted, it is probable that in Eden he was surrounded by a group of inoffensive animals, and that those creatures which he had cause to dread would have disappeared as he extended his dominion.　In this way the law of violent death and destruction which prevailed under the dynasties of the fish, the reptile, and the carnivorous mammifer would ultimately have been abrogated ; and under the milder sway of man life and peace would have reigned in a manner to which our knowledge of pre-Adamite and present nature may afford no adequate key. Be this as it may, on the important point of the original prevalence of death among the lower animals both records are at one.

8. In the department of "final causes," as they have been termed, Scripture and geology unite in affording large and interesting views.　They illustrate the procedure of the All-wise Creator during a long succession of ages, and thus enable us to see the effects of any of his laws, not only at one time, but in far distant periods.　To reject the consideration of this peculiarity of geological science would be the extremest folly, and would involve at once a misinterpretation of the geologic record and a denial of the agency of an intelligent Designer as revealed in Scripture, and indicated by the succession of beings.　Many of the past changes of the earth acquire their full significance only when taken in connection with the present wants of the earth's inhabitants ; and along the whole course of the geological history the creatures that we meet with are equally rich in the evidences of nice adaptation to circumstances and wonderful contrivances for special ends, with their modern representatives.　As an example of the former, how wonderful is the connection of the great

vegetable accumulations of the ancient coal swamps, and the bands and nodules of iron-stone which were separated from the ferruginous sands or clays in their vicinity by the action of this very vegetable matter, with the whole fabric of modern civilization, and especially with the prosperity of that race which, in our time, stands in the front of the world's progress. In a very ancient period, wide swamps and deltas, teeming with vegetable life, and which, if they now existed, would be but pestilent breeders of miasmata, spread over large tracts of the northern hemisphere, on which marine animals had previously accumulated thick sheets of limestone. Vast beds of vegetable matter were collected by growth in these swamps, and the waste particles that passed off in the form of organic acids were employed in concentrating the oxide of iron in underlying clays and sands. In the lapse of ages the whole of these accumulations were buried deep in the crust of the earth ; and long periods succeeded, when the earth was tenanted by reptilian and other creatures, unconscious of the treasures beneath them. The modern period arrived. The equable climate of the coal era had passed away. Continents were prepared for the residence of man, and the edges of the old carboniferous beds were exposed by subterranean movements, and laid bare by denudation. Man was introduced, fell from his state of innocence, and was condemned to earn his subsistence by the sweat of his brow ; and now for the first time appears the use of these buried coal swamps. They now afford at once the materials of improvement in the arts and of comfortable subsistence in extreme climates, and subjects of surpassing interest to the naturalist. Similar instances may be gleaned by the natural theologian from nearly every part of the geological history.

Lastly. Both records represent man as the last of God's works, and the culminating-point of the whole creation. We have already had occasion to refer to this as a result of zoology, geology, and Scriptural exegesis, and may here confine ourselves to the moral consequences of this great truth. Man is the capital of the column ; and, if marred and defaced by moral evil, the symmetry of the whole is to be restored, not by rejecting him altogether, like the extinct species of the ancient world, and replacing him by another, but by re-casting him in the image of his Divine Redeemer. Man, though recently introduced, is to exist eternally. He is, in one or another state of being, to be witness of all future changes of the earth. He has before him the option of being one with his Maker, and sharing in a future glorious and finally renovated condition of our planet, or of sinking into endless degradation. Such is the great spiritual drama of man's fate to be acted out on the theatre of the world. Every human being must play his part in it, and the present must decide what that part shall be. The Bible bases these great foreshadowings of the future on its own peculiar evidence ; yet I may venture humbly to maintain that its harmony with natural science, as far as the latter can ascend, gives to the Word of God a pre-eminent claim on the attention of the naturalist. The Bible, unlike every other system of religious doctrine, fears no investigation or discussion. It courts these. "While science," says a modern divine,* "is fatal to superstition, it is fortification to a Scriptural faith. The Bible is the bravest of books. Coming from God, and conscious of nothing but God's truth, it awaits the progress of knowledge with calm security. It watches the antiquary ransacking among classic ruins, and

---

* Hamilton.

rejoices in every medal he discovers and every inscription he deciphers; for from that rusty coin or corroded marble it expects nothing but confirmations of its own veracity. In the unlocking of an Egyptian hieroglyphic or the unearthing of some implement it hails the resurrection of so many witnesses; and with sparkling elation it follows the botanist as he scales Mount Lebanon, or the zoologist as he makes acquaintance with the beasts of the Syrian desert; or the traveller as he stumbles on a long-lost Petra or Nineveh or Babylon. And from the march of time it fears no evil, but calmly abides the fulfilment of those prophecies and the forthcoming of those events with whose predicted story inspiration has already inscribed its page. It is not light but darkness which the Bible deprecates; and if men of piety were also men of science, and if men of science were to search the Scriptures, there would be more faith in the earth, and also more philosophy."

The reader has, I trust, found in the preceding pages sufficient evidence that the Bible has nothing to dread from the revelations of geology, but much to hope in the way of elucidation of its meaning and confirmation of its truth. If convinced of this, I trust that he will allow me now to ask for the warnings, promises, and predictions of the Book of God his entire confidence; and, in conclusion, to direct his attention to the glorious prospects which it holds forth to the human race, and to every individual of it who, in humility and self-renunciation, casts himself in faith on that Divine Redeemer who is at once the creator of the heavens and the earth, and the brother and the friend of the penitent and the contrite. That same old book, which carries back our view to those ancient conditions of our planet which preceded not only the creation of man, but the earliest periods of which

science has cognizance, likewise carries our minds forward
into the farthest depths of futurity, and shows that all present
things must pass away.   It reveals to us a new heaven and
a new earth, which are to replace those now existing ; when
the Eternal Son of God, the manifestation of the Father
equally in creation and redemption, shall come forth conquering and to conquer, and shall sweep away into utter extinction all the blood-stained tyrannies of the present earth,
even as he has swept away the brute dynasties of the pre-
Adamite world, and shall establish a reign of peace, of love,
and of holiness that shall never pass away : when the purified
sons of Adam, rejoicing in immortal youth and happiness,
shall be able to look back with enlarged understandings and
grateful hearts on the whole history of creation and redemption, and shall join their angelic brethren in the final and
more ecstatic repetition of that hymn of praise with which
the heavenly hosts greeted the birth of our planet.   May God
in his mercy grant that he who writes and they who read
may "stand in their lot at the end of the days" and enjoy
the full fruition of these glorious prospects.

APPENDIX.

Q

APPENDIX.

A.—TRUE AND FALSE EVOLUTION.

THE term "evolution" need not in itself be a bugbear on theological grounds. The Bible writers would, I presume, have no objection to it if understood to mean the development of the plans of the Creator in nature. That kind of evolution to which they would object, and to which enlightened reason also objects, is the spontaneous evolution of nothing into atoms and force, and of these into all the wonderful and complicated plan of nature, without any guiding mind. Farther, biological and palæontological science, as well as the Bible, object to the derivation of living things from dead matter by merely natural means, because this can not be proved to be possible, and to the production of the series of organic forms found as fossils in the rocks of the earth by the process of struggle for existence and survival of the fittest, because this does not suffice to account for the complex phenomena presented by this succession. With reference to the testimony of palæontology, I have in other publications developed this very fully; and would here merely quote the summing up of the argument, as given in my Address of 1875 before the American Association for the Advancement of Science:

" I have thus far said nothing of the bearing of the prevalent ideas of descent with modification on this wonderful procession of life. None of these of course can be expected to take us back to the origin of living beings; but they also

fail to explain why so vast numbers of highly organized species struggle into existence simultaneously in one age and
disappear in another; why no continuous chain of succession
in time can be found gradually blending species into each
other; and why in the natural succession of things degradation under the influence of external conditions and final extinction seem to be laws of organic existence. It is useless
here to appeal to the imperfection of the record or to the
movements or migrations of species. The record is now in
many important parts too complete, and the simultaneousness
of the entrance of the faunas and floras too certainly established, and moving species from place to place only evades
the difficulty. The truth is that such hypotheses are at present premature, and that we require to have larger collections
of facts. Independently of this, however, it appears to me
that from a philosophical point of view it is extremely probable that all theories of evolution as at present applied to
life are fundamentally defective in being too partial in their
character ; and perhaps I can not better group the remainder
of the facts to which I wish to refer than by using them to illustrate this feature of most of the later attempts at generalization on this subject.

" First, then, these hypotheses are too partial in their tendency to refer numerous and complex phenomena to one
cause, or to a few causes only, when all trustworthy analogy
would indicate that they must result from many concurrent
forces and determinations of force. We have all no doubt
read those ingenious, not to say amusing, speculations in
which some entomologists and botanists have indulged with
reference to the mutual relations of flowers and haustellate
insects. Geologically the facts oblige us to begin with cryptogamous plants and mandibulate insects, and out of the desire of insects for non-existent honey, and the adaptations of
plants to the requirements of non-existent suctorial apparatus,
we have to evolve the marvellous complexity of floral form

and coloring, and the exquisitely delicate apparatus of the mouths of haustellate insects. Now when it is borne in mind that this theory implies a mental confusion on our part precisely similar to that which in the department of mechanics actuates the seekers for perpetual motion, that we have not the smallest tittle of evidence that the changes required have actually occurred in any one case, and that the thousands of other structures and relations of the plant and the insect have to be worked out by a series of concurrent evolutions so complex and absolutely incalculable in the aggregate that the cycles and epicycles of the Ptolemaic astronomy were child's play in comparison, we need not wonder that the common-sense of mankind revolts against such fancies, and that we are accused of attempting to construct the universe by methods that would baffle Omnipotence itself, because they are simply absurd. In this aspect of them indeed such speculations are necessarily futile, because no mind can grasp all the complexities of even any one case, and it is useless to follow out an imaginary line of development which unexplained facts must contradict at every step. This is also no doubt the reason why all recent attempts at constructing 'Phylogenies' are so changeable, and why no two experts can agree about the details of any of them.

"A second aspect in which such speculations are too partial is in the unwarranted use which they make of analogy. It is not unusual to find such analogies as that between the embryonic development of the individual animal and the succession of animals in geological time placed on a level with that reasoning from analogy by which geologists apply modern causes to explain geological formations. No claim could be more unfounded. When the geologist studies ancient limestones built up of the remains of corals, and then applies the phenomena of modern coral reefs to explain their origin, he brings the latter to bear on the former by an analogy which includes not merely the apparent results, but the causes at work, and the conditions of their action, and it is on this that the validity of his comparison depends, in so far as it relates to similarity of mode of formation. But when we compare the development of an animal from an embryo cell with the progress of animals in time, though we have a curious analogy as to the steps of the process, the conditions and causes at work are known to be altogether dissimilar, and therefore we have no evidence whatever as to identity of cause, and our reasoning becomes at once the most transparent of fallacies. Farther, we have no right here to overlook the fact that the conditions of the embryo are determined by those of a previous adult, and that no sooner does this hereditary potentiality produce a new adult animal than the terrible external agencies of the physical world, in presence of which all life exists, begin to tell on the organism, and after a struggle of longer or shorter duration it succumbs to death, and its substance returns into inorganic nature—a law from which even the longer life of the species does not seem to exempt it. All this is so plain and manifest that it is extraordinary that evolutionists will continue to use such partial and imperfect arguments. Another example may be taken from that application of the doctrine of natural selection to explain the introduction of species in geological time, which is so elaborately discussed by Sir C. Lyell in the last edition of his ' Principles of Geology.' The great geologist evidently leans strongly to the theory, and claims for it the ' highest degree of probability ;' yet he perceives that there is a serious gap in it, since no modern fact has ever proved the origin of a new species by modification. Such a gap, if it existed in those grand analogies by which we explain geological formations through modern causes, would be admitted to be fatal.

" A third illustration of the partial character of these hypotheses may be taken from the use made of the theory deduced from modern physical discoveries, that life must be merely a product of the continuous operation of physical laws. The

assumption, for it is nothing more, that the phenomena of life
are produced merely by some arrangement of physical forces,
even if it be admitted to be true, gives only a partial explanation of the possible origin of life. It does not account for the
fact that life as a force or combination of forces is set in antagonism to all other forces. It does not account for the
marvellous connection of life with organization. It does not
account for the determination and arrangement of forces implied in life. A very simple illustration may make this plain.
If the problem to be solved were the origin of the mariner's
compass, one might assert that it is wholly a physical arrangement both as to matter and force. Another might assert that
it involves mind and intelligence in addition. In some sense
both would be right. The properties of magnetic force and
of iron or steel are purely physical, and it might even be within the bounds of possibility that somewhere in the universe a
mass of natural loadstone may have been so balanced as to
swing in harmony with the earth's magnetism. Yet we would
surely be regarded as very credulous if we could be induced
to believe that the mariner's compass has originated in that
way. This argument applies with a thousandfold greater
force to the origin of life, which involves even in its simplest
forms so many more adjustments of force and so much more
complex machinery.

" Fourthly, these hypotheses are partial, inasmuch as they
fail to account for the vastly varied and correlated interde-
pendencies of natural things and forces, and for the unity of
plan which pervades the whole. These can be explained
only by taking into the account another element from without.
Even when it professes to admit the existence of a God, the
evolutionist reasoning of our day contents itself altogether
with the physical or visible universe, and leaves entirely out
of sight the power of the unseen and spiritual, as if this were
something with which science has nothing to do, but which
belongs only to imagination or sentiment. So much has this

been the case, that·when recently a few physicists and naturalists have turned to this aspect of the case, they have
seemed to be teaching new and startling truths, though only
reviving some of the oldest and most permanent ideas of our
race.   From the dawn of human thought it has been the conclusion alike of philosophers, theologians, and the common-
sense of mankind that the seen can be explained only by reference to the unseen, and that any merely physical theory of the
world is necessarily partial.   This, too, is the position of our
sacred Scriptures, and is broadly stated in their opening verse;
and indeed it lies alike at the basis of all true religion and all
sound philosophy, for it must necessarily be that ' the things
that are seen are temporal, the things that are unseen eternal.' With reference to the primal aggregation of energy in
the visible universe, with reference to the introduction of life,
with reference to the soul of man, with reference to the heavenly gifts of genius and prophecy, with reference to the introduction of the Saviour himself into the world, and with reference to the spiritual gifts and graces of God's people—all
these spring not from sporadic acts of intervention, but from
the continuous action of God and the unseen world, and this
we must never forget is the true ideal of creation in Scripture
and in sound theology.   Only in such exceptional and little influential philosophies as that of Democritus, and in the speculations of a few men carried off their balance by the brilliant
physical discoveries of our age, has this necessarily partial
and imperfect view been adopted.   Never, indeed, was its imperfection more clear than in the light of modern science.

  " Geology, by tracing back all present things to their origin,
was the first science to establish on a basis of observed facts
the necessity of a beginning and end of the world.   But even
physical science now teaches us that the visible world is a
vast machine for the dissipation of energy; that the processes
going on in it must have had a beginning in time, and that all
things tend to a final and helpless equilibrium.   This neces-

sity implies an unseen power, an invisible universe, in which the visible universe must have originated, and to which its energy is ever returning. The hiatus between the seen and the unseen may be bridged over by the conceptions of atomic vortices of force, and by the universal and continuous ether; but whether or not, it has become clear that the conception of the unseen as existing has become necessary to our belief in the possible existence of the physical universe itself, even without taking life into the account.

"It is in the domain of life, however, that this necessity becomes most apparent; and it is in the plant that we first clearly perceive a visible testimony to that unseen which is the counterpart of the seen. Life in the plant opposes the outward rush of force in our system, arrests a part of it on its way, fixes it as potential energy, and thus, forming a mere eddy, so to speak, in the process of dissipation of energy, it accumulates that on which animal life and man himself may subsist, and asserts for a time supremacy over the seen and temporal on behalf of the unseen and eternal. I say for a time, because life is, in the visible universe, as at present constituted, but a temporary exception, introduced from that unseen world where it is no longer the exception, but the eternal rule. In a still higher sense, then, than that in which matter and force testify to a Creator, organization and life, whether in the plant, the animal, or man, bear the same testimony, and exist as outposts put forth in the succession of ages from that higher heaven that surrounds the visible universe. In them, too, Almighty power is no doubt conditioned or limited by law, yet they bear more distinctly upon them the impress of their Maker; and, while all explanations of the physical universe which refuse to recognize its spiritual and unseen origin must necessarily be partial and in the end incomprehensible, this destiny falls more quickly and surely on the attempt to account for life and its succession on merely materialistic principles.

Q 2

"Here again, however, I must remind you that creation, as maintained against such materialistic evolution, whether by theology, philosophy, or Holy Scripture, is necessarily a continuous, nay, an eternal influence, not an intervention of disconnected acts. It is the true continuity, which includes and binds together all other continuity.

"It is here that natural science meets with theology, not as an antagonist, but as a friend and ally in its time of greatest need; and I must here record my belief that neither men of science nor theologians have a right to separate what God in Holy Scripture has joined together, or to build up a wall between nature and religion, and write upon it 'no thoroughfare.' The science that does this must be impotent to explain nature, and without hold on the higher sentiments of man. The theology that does this must sink into mere superstition.

"In conclusion, can we formulate a few of the general laws, or perhaps I had better call them general conclusions, respecting life, in which all palæontologists may agree? Perhaps it is not possible to do this at present satisfactorily, but the attempt may do no harm. We may, then, I think, make the following affirmations:

"1. The existence of life and organization on the earth is not eternal, nor even coeval with the beginning of the physical universe, but may possibly date from Laurentian or immediately pre-Laurentian times.

"2. The introduction of new species of animals and plants has been a continuous process, not necessarily in the sense of derivation of one species from another, but in the higher sense of the continued operation of the cause or causes which introduced life at first. This, as already stated, I take to be the true theological or Scriptural as well as scientific idea of what we ordinarily and somewhat loosely term creation.

"3. Though thus continuous, the process has not been uniform; but periods of rapid production of species have alternated with others in which many disappeared and few were introduced. This may have been an effect of physical cycles reacting on the progress of life.

"4. Species, like individuals, have greater energy and vitality in their younger stages, and rapidly assume all their varietal forms, and extend themselves as widely as external circumstances will permit. Like individuals also, they have their periods of old age and decay, though the life of some species has been of enormous duration in comparison with that of others; the difference appearing to be connected with degrees of adaptation to different conditions of life.

"5. Many allied species, constituting groups of animals and plants, have made their appearance at once in various parts of the earth, and these groups have obeyed the same laws with the individual and the species in culminating rapidly, and then slowly diminishing, though a large group once introduced has rarely disappeared altogether.

"6. Groups of species, as genera and orders, do not usually begin with their highest or lowest forms, but with intermediate and generalized types, and they show a capacity for both elevation and degradation in their subsequent history.

"7. The history of life presents a progress from the lower to the higher, and from the simpler to the more complex, and from the more generalized to the more specialized. In this progress new types are introduced and take the place of the older ones, which sink to a relatively subordinate place and become thus degraded. But the physical and organic changes have been so correlated and adjusted that life has not only always maintained its existence, but has been enabled to assume more complex forms, and that older forms have been made to prepare the way for newer, so that there has been on the whole a steady elevation culminating in man himself. Elevation and specialization have, however, been secured at the expense of vital energy and range of adaptation, until the new element of a rational and inventive nature was introduced in the case of man.

"8. In regard to the larger and more distinct types, we can not find evidence that they have, in their introduction, been preceded by similar forms connecting them with previous groups; but there is reason to believe that many supposed representative species in successive formations are really only races or varieties.

"9. In so far as we can trace their history, specific types are permanent in their characters from their introduction to their extinction, and their earlier varietal forms are similar to their later ones.

"10. Palæontology furnishes no direct evidence, perhaps never can furnish any, as to the actual transformation of one species into another, or as to the actual circumstances of creation of a species, but the drift of its testimony is to show that species come in *per saltum*, rather than by any slow and gradual process.

"11. The origin and history of life can not, any more than the origin and determination of matter and force, be explained on purely material grounds, but involve the consideration of power referable to the unseen and spiritual world.

"Different minds may state these principles in different ways, but I believe that, in so far as palæontology is concerned, in substance they must hold good, at least as steps to higher truths."

B.—EVOLUTION AND CREATION BY LAW.

EVOLUTIONIST writers have a great horror of what they term "intervention." But they should be informed that the idea of a planning Creator does not involve intervention in an extraordinary or miraculous sense, any more than what we call the ordinary operations of nature. It is a common but childish prejudice that every discovery of a secondary cause diminishes so much of what is to be referred to the agency of God. On the contrary, such discoveries merely aid us in comprehending the manner of his action. But when evolutionists, in their zeal to get rid of creative intervention, trace all things to the interaction of insensate causes, they fall into the absurdity of believing in absolute unmitigated chance as the cause of perfect order. Evidences of this may be found by the score in Darwin's works on the origin of species. I quote, however, from another and usually clear thinker, Wallace, in a review of the Duke of Argyll's "Reign of Law," which appeared some years ago, but represents very well this phase of thought :

"'It is curious,' says the Duke of Argyll, 'to observe the language which this most advanced disciple of pure naturalism [Mr. Darwin] instinctively uses, when he has to describe the complicated structure of this curious order of plants [the Orchids]. Caution in ascribing intentions to nature does not seem to occur to him as possible. Intention is the one thing which he does see, and which, when he does not see, he seeks for diligently until he finds it. He exhausts every form of words and of illustration by which intention or mental purpose can be described. 'Contrivance'—'curious contrivance'—'beautiful contrivance'—these are expressions which occur over and over again. Here is one sentence describing the parts of a particular species : 'the labellum is developed into a long nectary, *in order* to attract lepidoptera, and we shall presently give reason for suspecting that the nectar is *purposely* so lodged that it can be sucked only slowly, *in order* to give time for the curious chemical quality of this viscid matter setting hard and dry.'" Many other examples of similar expressions are quoted by the duke, who maintains that no explanation of these "contrivances" has been or can be given, except on the supposition of a personal contriver, specially arranging the details of each case, although causing them to be produced by the ordinary processes of growth and reproduction.

"Now there is a difficulty in this view of the origin of the structure of orchids which the duke does not allude to. The majority of flowering plants are fertilized, either without the agency of insects, or, when insects are required, without any very important modification of the structure of the flower. It is evident, therefore, that flowers might have been formed as varied, fantastic, and beautiful as the orchids, and yet have been fertilized by insects in the same manner as violets or clover or primroses, or a thousand other flowers. The strange springs and traps and pitfalls found in the flowers of orchids can not be necessary *per se*, since exactly the same end is gained in ten thousand other flowers which do not possess them. Is it not, then, an extraordinary idea to imagine the Creator of the universe *contriving* the various complicated parts of these flowers as a mechanic might contrive an ingenious toy or a difficult puzzle? Is it not a more worthy conception that they are some of the results of those general laws which were so co-ordinated at the first introduction of life upon the earth as to result necessarily in the utmost possible development of varied forms?"

A moment's thought is sufficient to show that there is no

essential difference between the Creator contriving every detail of the structure of an orchid and his producing it through some intermediate cause, or his commanding it into existence by his almighty word. The same mental process, so to speak, of the contriver is implied in either case. But there is an immeasurable difference between any of those ideas and that of the orchid producing its parts spontaneously under the operation of insensate physical law, whatever that may be, alone. Again, in the same review, Wallace writes:

"The uncertainty of opinion among naturalists as to which are species and which varieties is one of Mr. Darwin's very strong arguments that these two names can not belong to things quite distinct in nature and origin. The reviewer says that this argument is of no weight, because the works of man present exactly the same phenomena, and he instances patent inventions, and the excessive difficulty of determining whether they are new or old. I accept the analogy, and maintain that it is all in favor of Mr. Darwin's views; for are not all inventions of the same kind directly affiliated to a common ancestor. Are not improved steam-engines or clocks the lineal descendants of some existing steam-engine or clock? Is there ever a new creation in art or science any more than in nature? Did ever patentee absolutely originate any complete and entire invention no portion of which was derived from any thing that had been made or described before? It is, therefore, clear that the difficulty of distinguishing the various classes of inventions which claim to be new is of the same nature as the difficulty of distinguishing varieties and species, because neither are absolute new creations, but both are alike descendants of pre-existing forms, from which and from each other they differ by varying and often imperceptible degrees. It appears, then, that however plausible this writer's objections may seem, whenever he descends from generalities to any specific statement his supposed difficulties turn out to be in reality strongly confirmatory of Mr. Darwin's view."

Now that improved steam-engines are lineal descendants of other steam-engines is absolute nonsense, in any other aspect than that the structure of one suggested the structure of another to a contriving mind. We need not affirm this of God ; but we may affirm that the plans of the creative mind constitute the true link of connection between the different states and developments of inorganic and organic objects. This is the real meaning of creation by law, as distinguished from mere chance on the one hand, and arbitrary and capricious intervention on the other. Both of these extremes are equally illogical ; and it can not be too frequently repeated that divine revelation avoids both by maintaining with equal firmness the agency of the Creator, and that agency not capricious, but according to plan and purpose ; embracing not merely the action of the divine mind itself, but under it of all the forces and material things created.

C.—MODES OF CREATION.

A QUESTION often asked, but not easily answered, with reference to the creation of animals and plants, is—What was its precise method, and to what extent is such intervention conceivable. This is, it is true, not a properly scientific question, since science can not inform us of the act of creation. Nor is it properly a theological one, since revelation appeals to our faith in the facts, without giving us much information as to the mode. It can, therefore, be answered only conjecturally, except in so far as the law or plan of creation can be inferred from what is known, either from science or revelation, as to the history of life.

We may, in the first place, assume that law or plan must characterize creation. The Scriptural idea of it is not reconcilable with the supposition of a series of arbitrary acts any more than the scientific idea. The nature of these laws, as disclosed by Palæontology, has been already considered in a preceding part of this Appendix. What we may conjecture as to the nature of the creative act itself, from a comparison of nature and revelation, may be summed up as follows :

1. If we reduce organized beings to their ultimate organisms—cells or plastids—and with Spencer and Haeckel suppose these to be farther divisible into still smaller particles or plastidules, each composed of several complex particles of albumen or protoplasm, we may suppose the primary act of creation to consist in the aggregation of molecules of albuminous matter into such plastidules bearing the same relations, as "manufactured articles," to the future cell that inorganic molecules bear to crystals, and possessing within themselves

the potencies of organic forms. This is the nearest approach that we can make to the primary creative act, and its scientific basis is merely hypothetical, while revelation gives us no intimation as to any such constitution of organized matter.

2. The formulæ in Genesis, "Let the land produce," and "Let the waters produce," imply some sort of mediate creation through the agency of the land and the waters, but of what sort we have no means of knowing. They include, however, the idea of the origin of the lower and humbler forms of life from material pre-existing in inorganic nature, and also the idea of the previous preparation of the land and the waters for the sustenance of the creatures produced.

3. The expression in the case of man—" out of the dust "—would seem to intimate that the human body was constituted of merely elementary matter, without any previous preparation in organic forms. It may, however, be intended merely to inform us that, while the spirit is in the image of God, the bodily frame is "of the earth earthy," and in no respect different in general nature from that of the inferior animals.

4. The Bible indicates some ways in which creatures may be modified or changed into new species, or may give rise to new forms of life. The human body is, we are told, capable of transformation into a new or spiritual body, different in many important respects, and the future general prevalence of this change is an article of religious faith. The Bible represents the woman as produced from the man by a species of fission, not known to us as a natural possibility, except in some of the lower forms of life. The birth of the Saviour is represented as having been by parthenogenesis, and if it had pleased God that Jesus was to remain on earth as the progenitor of a new and higher type of man to replace that now existing, this might be regarded as the introduction of a new species. To what extent the Creator may have so acted on the constitution of organized beings as to produce changes of this kind we have no means of knowing; but if he have done so, we

may be sure that it has been in accordance with some definite plan or law.

5. We have a right to infer from Scripture that there must be some creative law which provides for the introduction of species, *de novo*, from unorganized matter, and which has been or is called into action by conditions as yet altogether un- known to us, and as yet inimitable, and therefore in some • sense miraculous. Whether we shall ever by scientific inves- tigation discover the law of this kind of divine intervention it is impossible to ˉsay. That all the theories of spontaneous generation and derivation hitherto promulgated are but wild guesses at it is but too evident.

6. Since in inorganic nature we meet with such ultimate facts as atoms of different kinds and with different properties, and ether of non-atomic constitution, all of which seem to be necessary to the existence of the world as it is, we may expect in like manner to find at the basis of organic structures and phenomena varied kinds of ultimate organisms and forces, probably much more complicated than those of inorganic nat- ure. The broad simplicity of existing theories of derivation and evolution is thus in itself a presumption against their truth, except as very partial explanations.

7. We have no right to consider the species " after their kinds" of revelation as coincident with the species recognized by science. Many of these may be merely races, the produc- tion of which in the course of time and in special circumstan- ces may fall within the powers of created species, and which may merely be the phases of such species in time and place. Only the accumulation of vast additional stores of facts can enable us to have any certain opinion on this point, and till it is settled the doctrine of derivation must remain purely hypothetical.

8. The inference of evolutionists that because certain forms of life succeed each other in geological time, they must have been derived from each other, has an aspect of truth and sim-

plicity ; but the idea of law or plan in creation suggests that the link of connection may be of a less direct nature than mere descent with modification. This has been referred to under a previous head.

9. In the scheme of revelation all the successions and changes of organized beings, just as much as their introduction at first, belong to the will and plan of God. Revelation opposes no obstacle to any scientific investigation of the nature and method of this plan, nor does it contemplate the idea that any discoveries of this kind in any way isolate the Creator from his works. Farther, inasmuch as God is always present in all his works, one part of his procedure can scarcely be considered an "intervention" any more than another.

10. As an illustration of the hypothetical condition of this subject, and of the views which may be taken as to its details, I quote from a memoir of my own certain conclusions with reference to the origin of the species of land plants which are found in the older geological formations. The conclusions stated are at the end of a detailed consideration of these plants and the circumstances of their occurrence :

"(1.) Some of the forms reckoned as specific in the Devonian and Carboniferous formations may be really derivative races. There are indications that such races may have originated in one or more of the following ways: (*a*) By a natural tendency in synthetic types to become specialized in the direction of one or other of their constituent elements. In this way such plants as *Arthrostigma* and *Psilophyton* may have assumed new varietal forms. (*b*) By embryonic retardation or acceleration,* whereby certain species may have had their maturity advanced or postponed, thus giving them various grades of perfection in reproduction and complexity of structure. The fact that so many Erian and Carboniferous plants seem to be on the confines of the groups of Acrogens

---

* In the manner illustrated by Hyatt and Cope.

and Gymnosperms may be supposed favorable to such ex-
changes. (*c*) The contraction and breaking up of floras which
occurred in the Middle Erian and Lower Carboniferous may
have been eminently favorable to the production of such varietal forms as would result from what has been called the
'struggle for existence.' (*d*) The elevation of a great expanse of new land at the close of the Middle Erian and the
beginning of the Coal period would, by permitting the extension of series over wide areas and fertile soils, and by removing the pressure previously existing, be eminently favorable
to the production of new, and especially of improved, varieties.

"(2.) Whatever importance we may attach to the above
supposed causes of change, we still require to account for the
origin of our specific types. This may forever elude our observation, but we may at least hope to ascertain the external
conditions favorable to their production. In order to attain
even to this it will be necessary to inquire critically, with reference to every acknowledged species, what its claims to distinctness are, so that we may be enabled to distinguish specific types from mere varieties. Having attained to some
certainty in this, we may be prepared to inquire whether the
conditions favorable to the appearance of new varieties were
also those favorable to the creation of new types, or the reverse—whether these conditions were those of compression
or expansion, or to what extent the appearance of new types
may be independent of any external conditions, other than
those absolutely necessary for their existence. I am not without hope that the further study of fossil plants may enable us
thus to approach to a comprehension of the laws of the creation, as distinguished from those of the continued existence
of species.

" In the present state of our knowledge we have no good
ground either to limit the number of specific types beyond
what a fair study of our material may warrant, or to infer that

such primitive types must necessarily have been of low grade, or that progress in varietal forms has always been upward. The occurrence of such an advanced and specialized type as that of *Syringoxylon* in the Middle Devonian should guard us against these errors. . The creative process may have been applicable to the highest as well as to the lowest forms, and subsequent deviations must have included degradation as well as elevation. I can conceive nothing more unreasonable than the statement sometimes made that it is illogical or even absurd to suppose that highly organized beings could have been produced except by derivation from previously existing organisms. This is begging the whole question at issue, depriving science of a noble department of inquiry on which it has as yet barely entered, and anticipating by unwarranted assertions conclusions which may perhaps suddenly dawn upon us through the inspiration of some great intellect, or may for generations to come baffle the united exertions of all the earnest promoters of natural science. Our present attitude should not be that of dogmatists, but that of patient workers content to labor for a harvest of grand generalizations which may not come till we have passed away, but which, if we are earnest and true to nature and its Creator, may reward even some of us." *

---

* Report on Fossil Plants of the Upper Silurian and Devonian, 1871.

D.—PRESENT CONDITION OF THEORIES OF LIFE.

ONE of the most learned and ingenious essays on this subject recently published* states on its first page that all the
varieties of opinion may be summed up under two heads :

" 1. Those which require the addition to ordinary matter of
an immaterial or spiritual essence, substance, or power, general or local, whose presence is the efficient cause of life ; and,

. " 2. Those which attribute the phenomena of life solely to the
mode of combination of the ordinary material elements of
which the organism is composed, without the addition of any
such immaterial essence, power, or force."

It is quite true that physiologists have up to this time
argued out these two alternatives, and that at present the
second is probably the more prevalent.   It is however also
true that neither includes or can possibly include the whole
truth, and that enlightened theism may enable us to hold
both, or all that is true in either.   Undoubtedly we must
hold that a higher spiritual power or Creator is necessary to
the existence of life ; but then this is necessary also to the
existence of dead matter and force.   So that if physiologists
think proper to trace the whole phenomena of life to material causes, they do not on that account in any way invalidate
the evidence for a spiritual Creator, nor for a spiritual element in the higher nature of man.   Yet so inconceivably shallow is much of the biological reasoning of the day, that it is
quite common to find physiologists referring all life to spontaneous and uncaused material agencies, because they have

---

* Drysdale's " Protoplasmic Theories of Life."

concluded that the arrangements of matter and force are sufficient to explain it ; and, on the other hand, to find theistic writers accusing physiology of materialism, if it finds the causes of vital phenomena in material forces, as if God could be present only in those processes which we can not understand.

What we really know as to the material basis of life may be summed up in a few words. Chemically, life is based on compounds of the albuminous group. These are highly complex in a molecular point of view, and seem to be formed in nature only where certain structures, those of the vegetable cell, exist under certain conditions. These albuminous substances do not necessarily possess vital properties. They may exist in a dead state just as other substances. Under certain conditions, however, those of forming part of a so-called living organism, they present phenomena of mechanical movement and molecular change, and of transformation or transmission of force, which enable them to transform themselves into various kinds of tissues, to nourish these when formed, and to establish a consensus of action between different parts of the organism ; and these properties are vastly varied in detail according to the kind of organism in which they take place, and the conditions under which the organism exists. The actually living matter presents no distinct structure recognizable by the microscope, and can not be distinguished chemically from ordinary albumen or protoplasm; but when living it must either exist in some peculiar and complex molecular arrangement unknown as yet to chemistry and physics, or must be actuated by some force or form of force called vital, and not as yet isolated or reduced to known laws or correlation. It does not concern theism or theology which of these may eventually prove to be the true view, or if it should be found, which is quite possible, that there is no real difference between them. In any case it is certain that in the lower animals, and in the merely physiological properties of man himself, living matter may act independently of any higher spiritual nature in the individual, though of course not independently of the higher power of God, which gave matter its properties and sustains them in their action. It is farther certain that in man the spiritual nature dominates and controls the vital, except when under abnormal conditions the latter unduly gains the mastery, and quenches altogether the spirit. In the language of the Bible, the merely vital endowments of the man belong to the flesh ($\sigma \acute{a} \rho \xi$), and to the rational mind or soul ($\psi \upsilon \chi \acute{\eta}$). The higher nature which man derives directly from God is the spirit ($\pi \nu \epsilon \tilde{\upsilon} \mu a$). Either of these parts of the complex humanity is capable of life ($\zeta \omega \acute{\eta}$) and of immortality. Either of them is capable of being in a state of death, though the import of this differs in its application to each. In Genesis, the body is composed of the ordinary earth-materials—the " dust of the ground." The higher nature is seen in the "shadow and likeness of God," and in the inbreathing of the Divine Spirit whereby man became a " living soul " in a higher sense than that in which the animals possess the ordinary "breath of life." With these views agree the later doctrines of the Bible as to the "trichotomy" of " body, soul, and spirit" in man, and of the added influence of the Spirit of God as acting on humanity.

R

E.—RECENT FACTS AS TO THE ORIGIN AND ANTIQUITY OF MAN.

SEVERAL recent statements as to new facts supposed to prove a pre-glacial antiquity for our species have been promulgated in scientific journals; but so great doubt rests upon them that they do not invalidate the statement that the earliest human remains belong to the post-glacial age. I may refer to the following:

A very remarkable discovery was made in 1875 by Professor Rutimeyer, of Basle. In a brown coal deposit of Tertiary, or at least of "inter-glacial" age—whatever that may mean in Switzerland—he found some fragments of wood so interlaced as to resemble wattle or basket-work. Steenstrup has, however, re-examined the evidence, and adduces strong reasons for the conclusion that the alleged human workmanship is really that of beavers.

The Swedish geologists have shown that there is no properly Palæolithic age in Scandinavia, and that even the reindeer had probably disappeared from Denmark and Sweden before their occupation by man. Some facts, however, seemed to indicate a residence of man in Sweden before the great post-pliocene subsidence. One of the most important of these is the celebrated hut of Sodertelge, referred to in this connection by Lyell. Recent observations have, however, shown that this hut was really covered by a landslip, and that its age may not be greater than eight centuries. Torel has recently explained this in the Proceedings of the Archæological Congress of Stockholm.

The human bone found in the Victoria Cave at Settle, apparently under a patch of boulder-clay, has been regarded as a good evidence of the pre-glacial origin of man. It has, however, always appeared to readers of the description as a very doubtful case; and Professor Hughes, of Cambridge, has recently expressed the opinion that the drift covering the bone may be merely a "pocket" of that material disengaged from a cavity in the limestone by the wearing of the cliff.

The same geologist has also shown reason to believe that the supposed case of the occurrence of palæolithic implements under boulder-clay near Brandon, discovered by Mr. Skertchley, and paraded by Geikie as a demonstration of the "inter-glacial" antiquity of man, in accordance with his system of successive glacial periods, is really an error, and has no foundation in the facts of the case.

Mr. Pengelly has endeavored to maintain the value of the deposit of stalagmite as a means of establishing dates, in his "Notes of Recent Notices of the Geology of Devonshire," Part I., 1874; but, I confess, with little success. He urges, in opposition to the Ingleborough Cave, that at Cheddar, where, according to him, no appreciable deposit whatever is taking place on the existing stalagmite. But this, of course, is evidence not applicable to the case in hand, as in the Cheddar case no stalagmite crust whatever would be produced. There are, no doubt, crevices and caves in which old stalagmite is even being removed or diminished in thickness. He farther asserts that in Kent's Cave teeth of the cave bear and other extinct animals are found covered by not more than an inch and a half of stalagmite, and consequently that if this were deposited at the rate of a quarter of an inch per annum—the supposed rate on the "Jockey Cap" at Ingleborough—these animals must have lived in Devonshire only six years ago, which is, of course, absurd. But he fails to perceive that this mode of occurrence is quite intelligible on the supposition of a rapid decrease in the amount of deposition in the later part of the stalagmite period. He farther refers to the fact that the thicker masses of stalagmite, which correspond to the places of more active drip of water, are in the same position in both crusts of stalagmite. This shows that the sources of water containing bicarbonate of lime have been the same from the first; but it proves nothing as to the rate of deposit.

Mr. Pengelly's own estimate of the rate of deposit gives, however, a length of time which is sufficient to show that there must be error somewhere in his calculations. He states the aggregate thickness of the two crusts at twelve feet, and then, assuming a rate of deposit of 0.05 inch in 250 years, or one inch in 5000 years, he arrives at the conclusion that the whole deposit required 720,000 years for its formation. He is "willing to suppose" the mechanical deposits to have accumulated more rapidly; but allowing one fourth of the time for them, we have nearly a million of years claimed for the residence of man in Devonshire, which, independently of other considerations, would push back the Palæozoic trilobites and corals of that county into the primitive reign of fire, and which in point of fact amounts to a *reductio ad absurdum* of the whole argument.

Professor Hughes* refers, as a case of rapid deposition of matter akin to stalagmite, to the deposit of travertine in the old Roman aqueduct of the Pont du Gard, near Avignon, where a thickness of fourteen inches seems to have accumulated in about 800 years. Mr. J. Carey has given in *Nature*, December 18, 1873, another instance where a deposit 0.75 inch thick was formed in fifteen years in a lead mine in Durham. Mr. W. B. Clarke in the same journal gives a case where in a cave at Brixton, known as Poole's Hole, a deposit one eighth of an inch in thickness was formed in six months. Such examples show how unsafe it is to reason as to the rate of deposit in by-gone times, and when climatal and local con-

---

* Lecture before the Royal Institution of London.

ditions may have been very different from those at present subsisting.

In an able address before the biological section of the British Association in 1876, Wallace adduces the following considerations as bearing on these questions ; and these are well worthy of attention as showing that it is the necessities of evolution rather than of geological facts that demand the assumption of a great antiquity for man, and induce so many writers to accept any evidence for this, however doubtful : (1) The great cerebral development of the so-called Palæolithic men, which shows no indications of graduating into inferior races. (2) The great variety of the implements of these ancient men, and the excellence of their carvings on bone and ivory, point to a similar conclusion. (3) Man is not related to any existing species of ape, but in various ways to several different species. (4) There is an accumulation of evidence to show that the earliest historical races excelled in many processes in the arts and in many kinds of culture. He instances the wonderful mechanical and engineering skill evidenced in the pyramids of Egypt in proof of this. His conclusion is either that the origin of man by development from apes must be pushed much farther back than any geologists at present hold, and I may add far beyond any probable date, or that he must have originated by some "distinct and higher agency" — which last is no doubt the true conclusion.

Haeckel, in his recent work, the "History of Creation," sketches the development of man from a monad, in twenty-two stages ; but he has to admit that stage twenty-first, or that of the "Ape-like man," nowhere exists, either recent or fossil. He has to assume that this missing link has perished in the submergence of an imaginary continent of Lemuria, in the Indian Ocean ; and it is instructive to observe that, after deducting this, his affiliation of the races of men, as indicated in a map of the distribution of the species, is in the main very similar to that with which we are familiar in ordinary collections of maps illustrative of the Bible.

The Post-glacial, Palæocosmic, or Palæolithic men of Europe are not improbably antediluvian ; and as to their precise date we know little.   As to postdiluvian man, Canon Rawlinson has recently pointed out* the remarkable convergence of all historic dates toward a time between 2000 to 3000 years B.C., or about the date of the Biblical deluge, which may reasonably be inferred to have occurred about 3200 B.C. He gives the following summary of historical origins as ascertained from the best data, and which accord with the representation of the Bible that in the time of Abraham the great monarchies of Egypt and the East were scarcely more powerful than the nomad tribe led by that patriarch :

Oldest date of Babylon			2300 B.C.
"	"	Assyria	1500
"	"	Iran	1500
"	"	India	1200
"	"	China	1154
"	"	Phœnicia	1700
"	"	Troad	2000
"	"	Egypt	2760
Sept. date of Deluge			3200

He rejects, of course, the fabulous chronologies of Egypt, China, and India as mythical, or referring to pre-human and antediluvian periods.   It is to be observed that while these dates place the origins of the oldest civilized nations at periods considerably subsequent to the deluge, they do not prevent us from supposing that these nations commenced their existence with an advanced civilization borrowed from antediluvian times, which is indeed a fair conclusion from the Biblical history, independently of the monumental evidence referred to by Wallace in a previous paragraph.

---

* *Leisure Hour,* 1876.

The Duke of Argyll, in his excellent little work " Primeval Man," in which he discusses the arguments in favor of primitive savagery advanced by Sir J. Lubbock in opposition to the views of Archbishop Whately in his lecture on the " Origin of Civilization," shows that there is no necessity to suppose a slow progress of mankind in the arts extending over indefinite ages ; and his argument in this respect connects itself with the facts as to the high cerebral organization of Palæocosmic men referred to above by Wallace. In summing up one division of his argument, he truly remarks : " If we assume with the supporters of the savage-theory that man has himself invented all that he now knows, then the very earliest inventions of our race must have been the most wonderful of all, and the richest in the fruits they bore. The man who first discovered the use of fire, and the use of those grasses which we now know under the name of corn, were discoverers compared with whom, as regards the value of their ideas to the world, Faraday and Wheatstone are but the inventors of ingenious toys. It may possibly be true, as Whately argues, that man never could have discovered these things without divine instruction. If so, it is fatal to the savage theory. But it is equally fatal to that theory if we assume the opposite position, and suppose that the noblest discoveries ever made by man were made by him in primeval times."

I may add that this is true, however far into antiquity we may stretch back these primeval times.

Professor E. S. Morse, in his address to the American Association, in 1876, as vice-president, takes as a theme the contributions of American zoologists to theories of evolution, and closes with those which refer to what he modestly terms " man's lowly origin." These contributions he sums up under three heads, as bearing on the following points: " 1. That in his earlier stages he reveals certain persistent characters of the ape ; 2. That the more ancient men reveal more ape-like features than the present existing men ; and, 3. That certain characteristics pertaining to early men still persist in the inferior races of men." Under the first head he gives contributions to the well-known fact that embryonic stages of the human being, like those of other high types, approximate to forms permanent in lower types. This is a fact inseparable from the law of reproduction; and as has been already shown in the text, absolutely without logical significance as even an analogical argument in favor of evolution. Under the second and third heads, he refers to cases of exceptional skulls and bones belonging to idiots and degraded races of men, as showing tendencies to lower forms, which as a matter of course they do, though with essential differences still marking them as human; and he assumes without any proof that these were relatively more common in primitive times, and that they are cases of reversion to a previous simian stage, instead of being results of abnormal conditions in the individual or variety. He sums up these arguments in the following paragraph:

" If we take into account the rapidly accumulating data of European naturalists concerning primitive man, with the mass of evidence presented in these notes, we find an array of facts which irresistibly point to a common origin with animals directly below us, and these evidences are found in the massive skulls with coarse ridges for muscular attachments, the rounding of the base of the nostrils, the early ossification of the nasal bones, the small cranial capacity in certain forms, the prominence of the frontal crest, the posterior position of the *foramen magnum*, the approximation of the temporal ridges, the lateral flattening of the tibia, the perforation of the humerus, the tendency of the pelvis to depart from its usual proportions; and, associated with all these, a rudeness of culture and the evidence of the manifestation of the coarsest instincts. He must be blind, indeed, who can not recognize the bearing of such grave and suggestive modifications."

Yet Professor Morse knows that there is no true specific or

even generic kinship between man and any species of ape; that the phenomena of idiocy and degeneracy have no real resemblance to those of distinct specific types; that the resemblances of man to apes, such as they are, point not in a direct manner to any stock of apes, but in a desultory way to several; and consequently that, if derived from any such animals, it must be from some stock altogether unknown to us as yet, either among recent or fossil animals. Farther, as Cope, himself an evolutionist, admits, while we can trace the skeletons of Eocene mammals through several directions of specialization in succeeding Tertiary times, man presents the phenomenon of an unspecialized skeleton which can not fairly be connected with any of these lines. Lastly, his quotation from Fiske, with reference to the supposed effect of a protracted infancy to develop the moral characteristics of man, though accompanied with the usual unfair and unreasonable sneer (which a naturalist like Morse should have been ashamed to quote) against men "still capable of believing that the human race was created by miracle in a single day," is the feeblest possible attempt to bridge over the gap between the spiritual nature of man and the merely psychical nature of brutes.

It is plain that if American naturalists have done nothing more in favor of the lowly origin of man than that which Professor Morse has been able, evidently with much industry and pains, to gather, we need not for the present abandon our claims to a higher origin. It is farther significant in connection with this that Professor Huxley, in his lectures in New York, while resting his case as to the lower animals mainly on the supposed genealogy of the horse, which has often been shown to amount to no certain evidence,* avoided altogether the discussion of the origin of man from apes, now obviously complicated with so many difficulties that both

---

* See critique in *International Review,* January, 1877.

Wallace and Mivart are staggered by them. Professor Thomas, in his recent lectures,* admits that there is no lower man known than the Australian, and that there is no known link of connection with the monkeys; and Haeckel † has to admit that the penultimate link in his phylogeny, the ape-like man, is absolutely unknown.

In Chapter XIII. I have not touched on the question of the absolute origin of language—this not being necessary to my argument. On this interesting subject, however, we have, in the naming of the animals by the first man, recorded in the second chapter of Genesis, not only the primary truth of his superiority to them, but a farther indication that the roots of human speech, other than interjectional, lie in onomatopœia, and especially in the voices of animals, and that the gift of speech was not the slow growth of ages, but an endowment of man from the first, just as much as any of his other powers or properties. An interesting discussion of this subject will be found in the concluding chapters of Wilson's " Pre-historic Man," second edition. Farther, the so-called " tallies " found with the bones of Palæocosmic men in European caves, and illustrated in the admirable work of Christy and Lartet, show that the rudiments even of writing were already in possession of the oldest race of men known to archæology or geology. (See Wilson, *op. cit.*, vol. ii., p. 54.)

I have not noticed, except incidentally, the alleged discoveries of very ancient human remains in America, as they all appear very problematical. There is, however, some evidence of the coexistence of man with the mastodon and other post-glacial animals in Illinois and elsewhere.

---

* Reported in *Nature*, 1876. † " History of Creation."

F.—BEARING OF GLACIAL PERIODS UPON THE INTERPRETATION OF GENESIS.

WHATEVER views may be taken as to that period of cold which occurs at the close of the Tertiary and beginning of the Modern period, it can not be held to have constituted any such break as to be considered, as it was at one time, an equivalent for the Biblical chaos. This is proved by the survival through this period of a very large proportion of the animals and plants still existing in the northern hemisphere. The chronological system of animals and plants has been continuous, as the Bible represents it, since their first appearance on earth.

It is further remarkable that while there is geological evidence of climates colder than the present in the temperate regions, there is equally good proof of warmer climates even within the arctic circle than those of the cold temperate regions at present. It is difficult to account for these vicissitudes of climate, and much controversy exists on the subject; but it seems certain that in the earlier Tertiary and Cretaceous periods, for example, the supplies of heat and light were so diffused over the earth as to permit the growth of a temperate vegetation in Greenland, and even in Spitzbergen. Geologists, however unwillingly, have been obliged to admit this as one of those great possibilities, altogether unexpected beforehand, which have been developed in the history of our planet. Various modes of explaining this succession of cold and warm periods have been adopted, all more or less hypothetical. Lyell has argued that it may be explained by a different distribution of land and water and of the ocean currents.

Croll accounts for it by the varying eccentricity of the earth's orbit, in connection with the precession of the equinoxes. Evans by a shifting of the axis of rotation of the earth. Drayson, Bell, Warring, and others, by a change in the inclination of the earth's axis. Others by the secular diminution of the internal heat of the earth, and of that of the sun. Others by the supposed recurrence of periods in which the sun gives more or less heat, or in which the earth is passing through colder or warmer regions of space. As the subject is of interest with reference to possible correspondences of these great summers and winters of the earth with the stages of the creative work, it may be well to notice shortly the relative merits of these theories.

(1.) The hypothesis of Croll is one of the most ingenious and elaborate of the whole; but it has two great defects. One is that the causes alleged are so uncertain and so complicated that it is difficult to estimate their real value. Another is that it proves too much, namely, a regular succession of cold and warm periods throughout geological time, of which we have no good evidence, and which is on many grounds improbable.

(2.) That the earth's axis of rotation has continued unchanged throughout the whole of the geological ages seems proved by the fact that the principal lines of crumpling and upheaval from the Laurentian period downward are arranged in great circles of the earth tangent to the polar circle; and that the lines of deposit of sediment in the Palæozoic age are coincident with the present direction of the arctic currents.

(3.) Astronomers consider it improbable that the obliquity of the ecliptic has materially changed, and serious differences of opinion exist as to the effects which a greater or less obliquity would produce on climate. It seems certain, however, that a less obliquity would occasion a more uniform distribution of heat and light throughout the year; and this, co-operating with other causes leading to a warm climate, might enable a temperate vegetation to approach the pole more closely than at present.

(4.) That the energy of the sun's radiation and the internal heat of the earth have been slowly decreasing seems certain ; but it is now generally admitted that these changes are so gradual that little effect can have been produced by them, except in the older geological periods, and that they can have no connection with the great glacial period of the Post-pliocene.

(5.) It is otherwise with the hypothesis that the sun's heat may, like that of some variable stars, have increased and diminished. There is, of course, no direct evidence of this, except the small differences observed in cycles of eleven and fifty-five years from the greater or less development of sunspots, and the analogy of observed variable stars. Still it is a possible cause of variations of climate. It might also aid in accounting for the extraordinary evidences of desert conditions and desiccation presented by the salt deposits of different geological periods in temperate latitudes.

(6.) The theory of the passage of the earth through zones of space of variable temperature is now generally abandoned, as there seems no reason to believe that such differences exist.

(7.) The theory of Lyell that changes in the distribution of land and water may, with the possible co-operation of other causes, have produced the observed diversities of climate, is that which seems best to meet the conditions presented. It is based on the known properties of land and water as to the absorption, radiation, and convection of heat, and on the remarkable diversities of climate in similar latitudes arising from this cause at present. Farther, it accords with the known fact that very great changes of level have occurred in connection with the glacial period. This theory undoubtedly embraces a true cause, admitted by all geologists, and it dispenses with the necessity of believing in the recurrence of glacial periods at regular intervals. It farther accords best with the evidence afforded by fossils, and especially by fossil plants. It has also the merit of directing due attention to the diversities of geographical conditions at different periods, and of dealing with causes of change operating within the earth itself. The only doubt with respect to it is its sufficiency to explain the changes which have occurred, and the view entertained of this will depend very much on the interpretation of the facts as to the intensity of the last glacial period. If moderate views can be taken of this, and if means can be found, by a less obliquity of the ecliptic or otherwise, to furnish a continuous supply of light in the arctic regions, the difficulties which have been alleged against it would disappear.

(8.) In connection with former periods of cold and warmth, and with the existence of temperate and tropical vegetation in polar latitudes, we should not forget that view which takes into account the probable effects of different conditions of the atmosphere, and the greater quantity of carbonic acid present in it, in early geological periods. This would, of course, best apply to the palæozoic floras, in so far as our present knowledge extends; but there may have been similar conditions in later periods. Dr. Sterry Hunt thus states this hypothesis:

" The agency of plants in purifying the primitive atmosphere was long since pointed out by Brongniart, and our great stores of fossil fuel have been derived from the decomposition, by the ancient vegetation, of the excess of carbonic acid of the early atmosphere, which through this agency was exchanged for oxygen gas. In this connection the vegetation of former periods presents the curious phenomenon of plants allied to those now growing beneath the tropics flourishing within the polar circles. Many ingenious hypotheses have been proposed to account for the warmer climate of earlier times, but are at best unsatisfactory, and it appears to me that the true solution of the problem may be found in the constitution of the early atmosphere, when considered in the light of Dr. Tyndall's beautiful researches on radiant heat. He has found that the presence of a few hundredths of carbonic-acid gas in the atmosphere, while offering almost no obstacle to the passage of the solar rays, would suffice to prevent almost entirely the loss by radiation of obscure heat, so that the surface of the land beneath such an atmosphere would become like a vast orchard-house, in which the conditions of climate necessary to a luxuriant vegetation would be extended even to the polar regions."

It is obvious that, in the production of complex effects of this kind, various causes, whether astronomical or connected with the mutations of the earth's crust, may have co-operated, and probably in all extreme cases did co-operate.

In any case it is evident that the vicissitudes of climate and the great pulsations of the crust, which have raised and depressed portions of the surface and changed the position of its covering of waters, have been potent agents in the hands of the Creator in effecting the changes and succession of living beings, which are thus, as Genesis intimates, children of the waters and of the land, and of the influences of the heavens. It is also interesting in this connection to observe that the occurrence of such periods of general warm climate as that in the Miocene shows that it would have been possible for man, under certain conditions, to have extended himself far more widely in his Edenic state than we can conceive of in the present condition of the earth. The modern world is perhaps even in this way "cursed" for man's sake.

G.—Dr. STERRY HUNT ON THE CHEMISTRY OF THE PRIMEVAL EARTH.

On looking back to the reference to this subject in Chapter V., I think it may be desirable to present to the reader in some more definite manner the conditions of a forming world; and I can not do this in any other way so well as by quoting the words of Dr. Sterry Hunt, as given in the abstract of his lecture on this subject delivered before the Royal Institution of London in 1867:

" This hypothesis of the nature of the sun and of the luminous process going on at its surface is the one lately put forward by Faye, and, although it has met with opposition, appears to be that which accords best with our present knowledge of the chemical and physical conditions of matter, such as we must suppose it to exist in the condensing gaseous mass which, according to the nebular hypothesis, should form the centre of our solar system. Taking this, as we have already done, for granted, it matters little whether we imagine the different planets to have been successively detached as rings during the rotation of the primal mass, as is generally conceived, or whether we admit with Chacornac a process of aggregation or concretion, operating within the primal nebular mass, resulting in the production of sun and planets. In either case we come to the conclusion that our earth must at one time have been in an intensely heated gaseous condition, such as the sun now presents, self-luminous, and with a process of condensation going on at first at the surface only, until by cooling it must have reached the point where the gaseous centre was exchanged for one of combined and liquefied matter.

"Here commences the chemistry of the earth, to the discussion of which the foregoing considerations have been only preliminary. So long as the gaseous condition of the earth lasted, we may suppose the whole mass to have been homogeneous; but when the temperature became so reduced that the existence of chemical compounds at the centre became possible, those which were most stable at the elevated temperature then prevailing would be first formed. Thus, for example, while compounds of oxygen with mercury or even with hydrogen could not exist, oxides of silicon, aluminium, calcium, magnesium, and iron might be formed and condense in a liquid form at the centre of the globe. By progressive cooling, still other elements would be removed from the gaseous mass, which would form the atmosphere of the non-gaseous nucleus. We may suppose an arrangement of the condensed matters at the centre according to their respective specific gravities, and thus the fact that the density of the earth as a whole is about twice the mean density of the matters which form its solid surface may be explained. Metallic or metalloidal compounds of elements, grouped differently from any compounds known to us, and far more dense, may exist in the centre of the earth.

"The process of combination and cooling having gone on until those elements which are not volatile in the heat of our ordinary furnaces were condensed into a liquid form, we may here inquire what would be the result, upon the mass, of a further reduction of temperature. It is generally assumed that in the cooling of a liquid globe of mineral matter, congelation would commence at the surface, as in the case of water; but water offers an exception to most other liquids, inasmuch as it is denser in the liquid than in the solid form. Hence ice floats on water, and freezing water becomes covered with a layer of ice, which protects the liquid below.

With most other matters, however, and notably with the various mineral and earthy compounds analogous to those which may be supposed to have formed the fiery-fluid earth, numerous and careful experiments show that the products of solidification are much denser than the liquid mass ; so that solidification would have commenced at the centre, whose temperature would thus be the congealing point of these liquid compounds. The important researches of Hopkins and Fairbairn on the influence of pressure in augmenting the melting-point of such compounds as contract in solidifying are to be considered in this connection.

" It is with the superficial portions of the ·fused mineral mass of the globe that we have now to do ; since there is no good reason for supposing that the deeply seated portions have intervened in any direct manner in the production of the rocks which form the superficial crust. This, at the time of its first solidification, presented probably an irregular, diversified surface from the result of contraction of the congealing mass, which at last formed a liquid bath of no great depth surrounding the solid nucleus. It is to the composition of this crust that we must direct our attention, since therein would be found all the elements (with the exception of such as were still in the gaseous form) now met with in the known rocks of the earth. This crust is now everywhere buried beneath its own ruins, and we can only from chemical considerations attempt to reconstruct it. If we consider the conditions through which it has passed, and the chemical affinities which must have come into play, we shall see that these are just what would now result if the solid land, sea, and air were made to react upon each other under the influence of intense heat. To the chemist it is at once evident that from this would result the conversion of all carbonates, chlorides, and sulphates into silicates, and the separation of the carbon, chlorine, and sulphur in the form of acid gases, which, with nitrogen, watery vapor, and a probable excess of oxygen, would

form the dense primeval atmosphere. The resulting fused
mass would contain all the bases as silicates, and must have
much resembled in composition certain furnace-slags or volcanic glasses. The atmosphere, charged with acid gases,
which surrounded this primitive rock must have been of immense density. Under the pressure of such a high baromet-
ric column, condensation would take place at a temperature
much above the present boiling-point of water, and the depressed portions of the half-cooled crust would be flooded
with a highly heated solution of hydrochloric acid, whose
action in decomposing the silicates is easily intelligible to
the chemist. The formation of chlorides of the various
bases, and the separation of silica, would go on until the affinities of the acid were satisfied, and there would be a separation of silica, taking the form of quartz, and the production
of a sea-water holding in solution, besides the chlorides of sodium, calcium, and magnesium, salts of aluminium and other
metallic bases. The atmosphere, being thus deprived of its
volatile chlorine and sulphur compounds, would approximate
to that of our own time, but differ in its greater amount of
carbonic acid.

"We next enter into the second phase in the action of the
atmosphere upon the earth's crust. This, unlike the first,
which was subaqueous, or operative only on the portion covered with the precipitated water, is sub-aerial, and consists in
the decomposition of the exposed parts of the primitive crust
under the influence of the carbonic acid and moisture of the
air, which convert the complex silicates of the crust into a
silicate of alumina, or clay, while the separated lime, mag-
nesia, and alkalies, being converted into carbonates, are carried down into the sea in a state of solution.

"The first effect of these dissolved carbonates would be to
precipitate the dissolved alumina and the heavy metals, after
which would result a decomposition of the chloride of calcium
of the sea-water, resulting in the production of carbonate of

lime or limestone, and chloride of sodium or common salt. This process is one still going on at the earth's surface, slowly breaking down and destroying the hardest rocks, and, aided by mechanical processes, transforming them into clays ; although the action, from the comparative rarity of carbonic acid in the atmosphere, is less energetic than in earlier times, when the abundance of this gas, and a higher temperature, favored the chemical decomposition of the rocks.   But now, as then, every clod of clay formed from the decay of a crystalline rock corresponded to an equivalent of carbonic acid abstracted from the atmosphere, and equivalents of carbonate of lime and common salt formed from the chloride of calcium of the sea-water."*

---

* See also Hunt, "Chemical and Geological Essays," p. 35.

H.—TANNIN AND BHEMAH.

THE following synopsis of the instances of the occurrence of the words *tannin* and *tan* will serve to show the propriety of the meaning, "great reptiles," assigned in the text to the former, as well as to illustrate the utility in such cases of "comparing Scripture with Scripture :"

1. TANNIN.

Exod. vii., 9.—Take thy rod and cast it before Pharaoh, and it shall become a *serpent*.

Probably a serpent, though perhaps a crocodile. (Septuagint, "δρά-κων.")

Deut. xxxii., 33. —Their vine is the poison of *dragons*.

Probably a species of serpent. (Septuagint, "δράκων.")

Job vii., 12.—Am I a sea, or a *whale*, that thou settest a watch over me.

Michaelis and others think, probably correctly, that the Nile and the crocodile, both objects of vigilance to the Egyptians, are intended. (Septuagint, "δράκων.")

Psa. lxxiv., 14.—Thou didst divide the sea by thy strength. Thou breakest the heads of the *dragons* in the waters.

Evidently refers to the destruction of the Egyptians in the Red Sea, under emblem of the crocodile. (Septuagint, "δράκων.")

Psa. xci., 13. — The young lion and the *dragon* thou shalt trample under foot.

The association shows that a powerful carnivorous animal is meant. (Septuagint, "δράκων.")

Psa. cxlviii., 7.—Praise the Lord, ye *dragons* and all deeps.

Evidently an aquatic creature. (Septuagint, "δράκων.")

Isa. xxvii., 1.—He shall slay the *dragon* in the midst of the sea [river].

A large predaceous aquatic animal (the crocodile), used here as an emblem of Egypt. (Septuagint, "δράκων.")

Isa. li., 9.—Hath cut Rahab and wounded the *dragon*.

Same as above.

Jer. li., 34. — [Nebuchadnezzar] hath swallowed me up as a *dragon.*
Ezek. xxix., 3.—Pharaoh, king of Egypt, the great *dragon* that lieth in the rivers.

A large predaceous animal. (Septuagint, "δράκων.")
In the Hebrew *tanim* appears by mistake for *tannin.* This is clearly the crocodile of the Nile. Verses 4 and 5 show that it is a large aquatic animal with *scales.* (Septuagint, "δράκων.")

2. TAN.

Psa. xliv., 19.—Thou hast sore broken us in the place of *dragons.*

Some understand this of shipwreck; but, more probably, the place of dragons is the desert. (Septuagint, "κάκωσις.")

Isa. xxxiv., 13.—[Bozrah in Idumea] shall be a habitation of *dragons* and a court of owls [or ostriches].
Isa. xliii., 20.—The wild beasts shall honor me, the *dragons* and the ostriches, because I give water in the wilderness.
Isa. xiii., 22.—*Dragons* in their pleasant palaces.

An animal inhabiting ruins, and associated with the ostrich. (Septuagint, "σειρήν.")

Evidently an animal of the dry deserts. (Septuagint, "σειρήν.")

Represented as inhabiting the ruins of Babylon, and associated with wild beasts of the desert. (Septuagint, "ἐχῖνος.")

Isa. xxxv., 7.—And the parched ground shall become a pool, and the thirsty land springs of water; in the habitation of *dragons,* where each lay, shall be grass with reeds and rushes.
Job xxx., 29.—I am a brother of *dragons* and a companion of ostriches.

An animal making its lair or nest in dry, parched places. (Septuagint, "ὄρνις.")

The association indicates an animal of the desert, and the context that its cry is mournful. (Septuagint, "σειρήν.")

Jer. ix., 11; x., 22.—I will make Jerusalem heaps, a den of *dragons.*

Same as above. See also Jeremiah xlix., 33; li., 37; and Mal. i., 3, where the word is in the female form (*tanoth*). (Septuagint, "δράκων" and "στρουθός.")

Lam. iv., 3.—Even the *sea-mon-sters* draw out the breast, they give suck to their young ones. The daughter of my people is become cruel, like the ostriches in the wilderness.

In the Hebrew text the word is *tannin*, evidently an error for *ta-nim*. The suckling of young, and association of ostriches, agree with this. (Septuagint, "ἐράκων.")

Micah i., 8.—I will make a wailing like the *dragons*, and mourning like the owls [ostriches].

The wailing cry accords with the view of Gesenius that the jackal is meant. (Septuagint, "δράκων.")

We learn from the above comparative view that the *tannin* is an aquatic animal of large size, and predaceous, clothed with scales, and a fit emblem of the monarchies of Egypt and Assyria. In two places it is possible that some species of serpent is denoted by it. We must suppose, therefore, that in Genesis i. it denotes large crocodilian and perhaps serpentiform reptiles. The *tan* is evidently a small mammal of the desert.

I omitted to notice in the text a criticism of my explanation of the word *bhemah* in "Archaia," made in Archdeacon Pratt's "Scripture and Science not at Variance" (edition of 1872). He opposes to the meaning of "herbivorous animals" which I have sought to establish, two exceptional passages. In one of these, Deut. xxviii., 26, the word is used in its most general sense for all beasts, which the context shows can not be its meaning in Gen. i. In the other, Prov. xxx., 30, he says it is applied to the lion. The actual expression used, however, merely implies that the lion is "mighty among *bhemah*," the comparison being probably between the strength of the lion and that of oxen, antelopes, and other strong and active creatures. It does not affirm that the lion is one of the *bhemah*. While I have every respect for the erudition of Archdeacon Pratt, and highly value his book, I must regard this objection as an example of a style of biblical exposition much to be deprecated, though too often employed.

I.—ANCIENT MYTHOLOGIES.

THE current views respecting the relations of ancient mythologies with each other and with the Bible have been continually shifting and oscillating between extremes. The latest and at present most popular of these extreme views is that so well expounded by Dr. Max Müller in his various essays on these subjects, and which traces at least the Indo-European theogony to a mere personification of natural objects. The views given in the text are those which to the author appear alone compatible with the Bible, and with the relations of Semitic and Aryan theology ; but, as the subject is generally regarded from a quite different point of view, a little further explanation may be necessary.

1. According to the Bible, spiritual monotheism is the primitive faith of man, and with this it ranks the doctrine of a malignant spirit or being opposed to God, and of a primitive state of perfection and happiness. It is scarcely necessary to say that these doctrines may be found as sub-strata in all the ancient theologies.

2. In the Hebrew theology the fall introduces the new doctrine of a mediator or deliverer, human and divine, and an external symbolism, that of the cherubic forms, composite figures made up of parts of the man, the lion, the ox, and the eagle. These forms are referred back to Eden, where they are manifestly the emblems of the perfections of the Deity, lost to man by the fall, and now opposed to his entrance into Eden and access to the tree of life, the symbol of his immortal happiness. Subsequently the cherubim are the visible indications of the presence of God in the tabernacle and temple ; and in

the Apocalypse they reappear as emblems of the Divine perfections, as reflected in the character of man redeemed. The cherubim, as guardians of the sacred tree, and of sacred places in general, appear in the worship of the Assyrians and Egyptians, as the winged lions and bulls of the former, and the sphinx of the latter. They can also be recognized in the sepulchral monuments of Greek Asia and of Etruria. Farther, it was evidently an easy step to proceed from these cherubic figures to the adoration of sacred animals. But the cherubic emblems were connected with the idea of a coming Redeemer, and this was with equal ease perverted into hero-worship. Every great conqueror, inventor, or reformer was thus recognized as in some sense the " coming man," just as Eve supposed she saw him in her first-born. In addition to this, the sacredness of the first mother as the mother of the promised seed of the woman, led to the introduction of female deities.

3. The earliest ecclesiastical system was the patriarchal, and this also admitted of corruption into idolatry. The great patriarch, venerable by age and wisdom, when he left this earth for the spirit world, was supposed there, in the presence of God, to be the special guardian of his children on earth. Some of the gods of Egypt and of Greece were obviously of this character, and in China and Polynesia we see at this day this kind of idolatry in a condition of active vitality.

4. As stated in the text, the mythology of Egypt and Greece bears evident marks of having personified certain cosmological facts akin to those of the Hebrew narrative of creation. In this way ancient idolators disposed of the prehistoric and pre-Adamite world, changing it into a period of gods and demigods. This is very apparent in the remarkable Assyrian Genesis recovered by the late George Smith from the clay tablets found in the ruined palace of Assurbanipal.

5. In all rude and imaginative nations, which have lost the distinct idea of the one God, the Creator, nature becomes

S

more or less a source of superstitions. Its grand and more rare phenomena of volcanoes, earthquakes, thunder-storms, eclipses, become supernatural portents ; and as the idea of power associates itself with them, they are personified as actual agents and become gods. In like manner, the more constant and useful objects and processes of nature become personified as beneficent deities. This may be, to a great extent, the character of the Aryan theology ; but, except where all ideas of primitive religion and traditions of early history have been lost, it can not be the whole of the religion of any people. The Bible negatively recognizes this source of idolatry, in so constantly referring all natural phenomena to the divine decree. In connection with this, it is worthy of remark that rude man tends to venerate the new animal forms of strange lands. Something of this kind has probably led some of the American Indians to give a sort of divine honor to the bear. It was in Egypt that man first became familiar with the strange and gigantic fauna of Africa, whose effect on his mind in primitive times we may gather from the book of Job. In Egypt, consequently, there must have been a strong natural tendency to the adoration of animals.

The above origins of idolatry and mythology, as stated or implied in the Bible, of course assume that the Semitic monotheistic religion is the primitive one. The first deviations from it probably originated in the family of Ham. A city of the Rephaim of Bashan was in the days of Abraham named after Ashtoreth Karnaim—the two-horned Astarte, a female divinity and prototype of Diana, and perhaps an historic personage, in whom both the moon and the domestic ox were rendered objects of worship. This is the earliest Bible notice of idolatry.* In Egypt a mythology of complex diversity existed at least as far back. We must remember, however, that Egypt is Cush as well as Mizraim, and its idolatry is probably

---

* Except, perhaps, Job xxxi., 27.

to be traced, in the first instance, to the Nimrodic empire, from which, as from a common centre, certain new and irreligious ideas seem to have been propagated among all the branches of the human family. It is quite probable that the correspondences between Egyptian, Greek, and Hindoo myths go back as far as to the time when the first despotism was erected on the plain of Shinar, and when able but ungodly men set themselves to erect new political and social institutions on the ruins of all that their fathers had held sacred. In addition to this, the mythology and language of the Aryans alike bear the impress of the innovating and restless spirit of the sons of Japhet.

I have stated the above propositions to show that the Bible affords a rational and connected theory of the origin of the false religions of antiquity ; and to suggest as inquiries in relation to every form of mythology—how much of it is primitive monotheism, how much cherub-worship, how much hero-worship, how much ancestor-worship, how much distorted cosmogony, how much pure idealism and superstition, since all these are usually present. I may be allowed further to remind the reader how much evidence we have, even in modern times, of the strong tendency of the human mind to fall into one or another of these forms of idolatry ; and to ask him to reflect that really the only effectual conservative element is that of revelation. How strong an argument is this for the necessity to man of an inspired rule of religious faith.

[The above note was in substance contained in the Appendix to "Archaia" in 1860, and its correctness has, I think, been confirmed by subsequent discoveries.]

K.—ASSYRIAN AND EGYPTIAN TEXTS.

PROGRESS is continually being made in the decipherment and publication of these, and new facts are coming to light in consequence as to the religions of the early postdiluvian period.

According to the late George Smith and to Mr. Sayce, in their contributions to Bagster's "Records of the Past," the earliest monumental history of Babylonia reveals two races, the Akkadian or Urdu, a Turanian race, with an agglutinate language of the Finnish or Tartar type, and the Sumir or Keen-gi, believed to be Shemitic. The race of Akkad seems to have invented the cuneiform writing at a very early period, and it no doubt represents the primitive Cushites of the Bible, to whom is attributed the empire of Nimrod, whose first cities were Babel and Erech and Akkad and Calneh. Very ancient inscriptions of this early Chaldean or Cushite race exist, probably earlier than the time of Abraham. That of king Urukh, who is called "a very ancient king," on an inscription of Nabonadius, 555 B.C., represents himself as building temples to several gods and goddesses, so that in his time there was already a developed polytheism, unless, indeed, he was himself the inventor or introducer of much of it. Yet one can gather from the probably contemporary Creation and Deluge tablets translated by Mr. Smith, that a Supreme God was still recognized, and that the subordinate deities, though their worship was probably gaining in importance, were still only local and created beings. Yet it was undoubtedly from this embryo idolatry that Abraham dissented, and was thus led to leave his native land.

In like manner, in the early Egyptian Hymn to Amen Ra, translated by Mr. Goodwin, though we have the gods mentioned, they are inferior beings, and not higher in position than the angels of the Old Testament, while Ra himself is " Lord of Eternity, Maker Everlasting," and is praised as

> "Chief creator of the whole earth,
> Supporter of affairs above every god,
> In whose goodness the gods rejoice."

Thus, although there can be little doubt that Ra was a sun-god, there can be as little that he is the Il or El of the Shemitic peoples, and that his worship represents that of the one God, the Creator. It seems probable also that there was an esoteric doctrine of this kind among the priests and the educated, however gross the polytheism of the vulgar. In short, the state of things in Assyria and Egypt was not dissimilar from that prevailing at this day in India, where learned men may fall back upon the ancient Vedas, and maintain that their religion is monotheistic, while the common people worship innumerable gods. All this points to a primitive monotheism, just as the peculiar forms of adoration given to saints and the Virgin Mary in the Greek and Roman churches historically imply a primitive Christianity on which these newer beliefs and rites have been engrafted.

L.—SPECIES AND VARIETAL FORMS WITH REFERENCE TO THE UNITY OF MAN.

In the concluding chapters of "Archaia" the nature of species, as distinguished from varieties, was discussed, and specially applied to the varieties and races of man. This discussion has been omitted from the text of the present work; but, in an abridged form, is introduced here, with especial reference to those more recent views of this subject now prevalent in consequence of the growth of the philosophy of evolution; but which I feel convinced must, with the progress of science, return nearer to the opinions held by me in 1860, and summarized below.

We can determine species only by the comparison of individuals. If all these agree in all their characters except those appertaining to sex, age, and other conditions of the individual merely, we say that.they belong to the same species. If all species were invariable to this extent, there could be no practical difficulty, except that of obtaining specimens for comparison. But in the case of very many species there are minor differences, not sufficient to establish specific diversity, but to suggest its possibility; and in such cases there is often great liability to error. In cases of this kind we have principally two criteria: first, the nature and amount of the differences; secondly, their shading gradually into each other, or the contrary. Under the first of these we inquire — Are they no greater in amount than those which may be observed in individuals of the same parentage? Are they no greater than those which occur in other species of similar structure or habits? Do they occur in points known in other species to be readily variable, or in points that usually remain unchanged? Are none of them constant in the one supposed species, and constantly absent in the other? 'Under the second we ask— Are the individuals presenting these differences connected together by others showing a series of gradations uniting the extremes by minute degrees of difference? If we can answer these questions—or such of them as we have the means of answering—in the affirmative, we have no hesitation in referring all to the same species. If obliged to answer all or many in the negative, we must at least hesitate in the identification; and if the material is abundant, and the distinguishing characters clear and well defined, we conclude that there is a specific difference.

Species determined in this way must possess certain general properties in common :

1. Their individuals must fall within a certain range of uniform characters, wider or narrower in the case of different species.

2. The intervals between species must be distinctly marked, and not slurred over by intermediate gradations.

3. The specific characters must be invariably transmitted from generation to generation, so that they remain equally distinct in their limits if traced backward or forward in time, in so far as our observation may extend.

4. Within the limits of the species there is more or less liability to variation; and this, though perhaps developed by external circumstances, is really inherent in the species, and must necessarily form a part of its proper description.

5. There is also a physiological distinction between species, namely, that the individuals are sterile with one another, whereas this does not apply to varieties ; and though Darwin has labored to break down this distinction by insisting on rare exceptional cases, and suggesting many supposed ways by which varieties of the same species might possibly attain to this

kind of distinctness, the difference still remains as a fact in nature; though one not readily available in practically distinguishing species.

These general properties of species will, I think, be admitted by all naturalists as based on nature, and absolutely necessary to the existence of natural history as a science, independently of any hypotheses as to the possible changes of specific forms in the lapse of time. I now proceed to give a similar summary of the laws of the varieties which may exist — always, be it observed, within the limits of the species.

1. The limits of variation are very different in different species. There are many in which no well-marked variations have been observed. There are others in which the variations are so marked that they have been divided, even by skilful naturalists, into distinct species or even genera. I do not here refer to differences of age and sex. These in many animals are so great that nothing but actual knowledge of the relation that subsists would prevent the individuals from being entirely separated from one another. I refer merely to the varieties that exist in adults of the same sex, including, however, those that depend on arrest of development, and thus make the adult of one variety resemble in some respects the young of another; as, for instance, in the hornless oxen, and beardless individuals among men. If we inquire as to the causes on which the greater or less disposition to vary depends, we must, in the first place, confess our ignorance, by saying that it appears to be in a great measure constitutional, or dependent on minute and as yet not distinctly appreciable structural, physiological, and psychical characters. Darwin states that Pallas long ago suggested, from the known facts that the seeds of hybrid plants and grafted trees are very variable, the theory that mixture of breeds tends to produce variability; but Darwin does not seem to attach much importance to this, and admits our inability to explain the origin of these differ-

ences.* We know, however, certain properties of species that are always or usually connected with great liability to variation. The principal of these are the following : 1. The liability to vary is, in many cases, not merely a specific peculiarity; it is often general in the members of a genus or family. Thus the cats, as a family, are little prone to vary; the wolves and foxes very much so. 2. Species that are very widely distributed over the earth's surface are usually very variable. In this case the capacity to vary probably adapts the creature to a great variety of circumstances, and so enables it to be widely distributed. It must be observed here that hardiness and variability of constitution are more important to extensive distribution than mere locomotive powers, for matters have evidently been so arranged in nature that, where the habitat is suitable, colonists will find their way to it, even in the face of difficulties almost insurmountable. 3. Constitutional liability to vary is sometimes connected with or dependent on extreme simplicity of structure, in other cases on a high degree of intelligence and consequent adaptation to various modes of subsistence. Those minute, simply organized, and very variable creatures, the Foraminifera, exemplify the first of these apparent causes; the crafty wolves furnish examples of the second. 4. Susceptibility to variation is farther modified by the greater or less adaptability of the digestive and locomotive organs to varied kinds of food and habitat. The monkeys, intelligent, imitative, and active, are nevertheless very limited in range and variability, because they can comfortably subsist only in forests, and in the warmer regions of the earth. The hog, more sluggish and less intelligent, has an omnivorous appetite, and no very special requirements of habitat, and so can vary greatly and extend over a large portion of the earth. Farther, in connection with this subject it may be observed that the conditions favorable to variation are also in the case

---

* . "Animals and Plants under Domestication," p. 406.

of the higher animals favorable to domestication, while it may
also be affirmed that, other things being equal, animals in a
domesticated state are much more liable to vary than those
in a wild state, and this independent of intentional selection.
Darwin admits this, and gives many examples of it.

2. Varieties may originate in two different ways. In the
case of wild animals it is generally supposed that they are
gradually induced by the slow operation of external influences;
but it is certain that in domesticated animals they often appear suddenly and unexpectedly, and are not on that account
at all less permanent. A large proportion of our breeds of
domestic animals appear to originate in this way. A very remarkable instance is that of the " Niata " cattle of the Banda
Orientale, described by Darwin in his " Voyage of a Naturalist."
These cattle are believed to have originated about a century
ago among the Indians to the south of the La Plata, and the
breed propagates itself with great constancy. " They appear,"
says Darwin, " externally to hold nearly the same relation to
other cattle which bull-dogs hold to other dogs. Their fore-
head is very short and broad, with the nasal end turned up,
and the upper lip much drawn back; their lower jaws project
outward; when walking they carry their heads low on a short
neck, and their hinder legs are rather longer compared with
the front legs than is usual." It is farther remarkable in respect to this breed that it is, from its conformation of head,
less adapted to the severe droughts of those regions than the
ordinary cattle, and can not, therefore, be regarded as an
adaptation to circumstances. In his later work on animals
under domestication, Darwin gives many other instances of
the origination of breeds of cattle and other animals in this
abrupt and mysterious manner, and without any selection,
though he strongly leans to the conclusion that slow and
gradual changes are the most frequent causes of variation.
It is to be observed, however, that very slow changes are in
more danger of being accidentally diverted or obliterated by

crossing, and that the first stages of an incipient change may be too unimportant to be permanent.

Many writers on the subject of the Unity of Man assume that any marked variety must require a long time for its production. Our experience in the case of the domestic animals teaches the reverse of this view; a very important point too often overlooked.

3. The duration or permanence of varieties is very different. Some return at once to the normal type when the causes of change are removed. Others perpetuate themselves nearly as invariably as species, and are named races. It is these races only that we are likely to mistake for true species, since here we have that permanent reproduction which is one of the characteristics of the species. The race, however, wants the other characteristics of species as above stated; and it differs essentially in having branched from a primitive species, and in not having an independent origin. It is quite evident that in the absence of historical evidence we must be very likely to err by supposing races to have really originated in distinct "primordial forms." Such error is especially likely to arise if we overlook the fact of the sudden origination of such races, and their great permanency if kept distinct. There are two facts which deserve especial notice, as removing some of the difficulty in such cases. One is that well-marked races usually originate only in domesticated animals, or in wild animals which, owing to accidental circumstances, are placed in abnormal circumstances. Another is, that there always remains a tendency to return, in favorable circumstances, to the original type. This tendency to reversion is much underrated by Darwin and his followers; yet they constantly recur to it as a means of proving possible derivation, and their writings abound in examples of it. Perhaps the most remarkable of these reversions are those which occur when varieties destitute of all the markings of the original stock are crossed and re-

produce those markings, which Darwin shows to occur in
pigeons and domestic fowls. The domesticated races usually require a certain amount of care to preserve them in a
state of purity, both on this account and on account of the
readiness with which they intermix with other varieties of the
same species. Many very interesting facts in illustration of
these points might be adduced. The domesticated hog differs in many important characters from the wild boar. In
South America and the West Indies it has returned, in three
centuries or less, to its original form.* The horse is probably not known in a state originally wild, but it has run wild in
America and in Siberia. In the prairies of North America,
according to Catlin,† they still show great varieties of color.
The same is the case in Sable Island, off the coast of Nova
Scotia,‡ where herds of wild horses have existed since an
early period in the settlement of America. In South America and Siberia they have assumed a uniform chestnut or bay
color. In the plains of Western America they retain the dimensions and vigor of the better breeds of domesticated
horses. In Sable Island they have already degenerated to
the level of Highland ponies; but in all countries where
they have run wild, the elongated and arched head, high
shoulders, straight back, and other structural characters probably of the original wild horse, have appeared. We also
learn from such instances that, while races among domesticated animals may appear suddenly, they revert to the original type, when unmixed, comparatively slowly; and this especially when the variation is in the nature of degeneracy.

4. Some characters are more subject to variation than others. In the higher animals variation takes place very readily

---

* Prichard. This is admitted by Darwin, who gives other examples,
though he insists much on the climatal variations which still remain in
feral pigs.
† "North American Indians."
‡ Haliburton's "Nova Scotia;" Gilpin's Lecture on Sable Island.

in the color and texture of the skin and its appendages. This, from its direct relation to the external world, and ready sympathy with the condition of the digestive organs, might be expected to take the lead. In those domesticated animals which are little liable to vary in other repects, as the cat and duck, the color very readily changes. Next may be placed the stature and external proportions, and the form of such appendages as the external ear and tail. All these characters are very variable in domestic animals. Next we may place the form of the skull, which, though little variable in the wild state, is nearly always changed by domestication. Psychological functions, as the so-called instincts of animals, are also very liable to change, and to have these changes perpetuated in races. Very remarkable instances of this have been collected by Sir C. Lyell * and Dr. Prichard. Lastly, important physiological characters, as the period of gestation, etc., and the structure of the internal organs connected with the functions of nutrition, respiration, etc., are little liable to change, and remain unaffected by the most extreme variations in other points; and it is, no doubt, in these more essential and internal parts that the tendency survives to return under favorable circumstances to the original type.

5. Varieties or races of the same species are fully reproductive with each other, which is not the case with true species. Mutual sterility of varieties of the same species is an exceptional peculiarity, if it ever truly exist ; and, on the other hand, the cross-fertilization of varieties of the same species, whether in animals or plants, tends to vigorous life, and also to return to the primitive or average type. On the other hand, intermixture of distinct species rarely, if ever, occurs freely in nature. It is generally a result of artificial

---

* " Principles of Geology ;" " Natural History of Man." See also a very able article on the " Varieties of Man," by Dr. Carpenter, in Todd's Cyclopædia.

contrivance. Again, hybrids produced from species known to be distinct are either wholly barren, or barren *inter se*, reproducing only with one of the original stocks, and rapidly returning to it; or if ever fertile *inter se*, which is somewhat doubtful, rapidly run out. It has been maintained by Pallas and others, and Darwin leans to this idea, that there is still another possibility, namely, that of the perfect and continued fertility of such mixed races, especially after long domestication; but their proofs are derived principally from the intermixture of the races of dogs and of poultry, which are cases actually in dispute at present, as to the original unity or diversity of the so-called species.

If we apply these considerations to man, our conclusion must be that, even in his bodily frame, he is not merely specifically but ordinally distinct from other animals, and that the differences between races of men are varietal rather than specific. This view is confirmed by the following facts:

1. The case of man is not that of a wild animal; and it presents many points of difference even from the case of the domesticated lower animals. According to the Bible history, man was originally fitted to subsist on fruits, to inhabit a temperate climate, and to be exempt from the necessity of destroying or contending with other animals. This view unquestionably accords very well with his organization. He still subsists principally on vegetable food, is most numerous in the warmer regions of the earth; and, when so subsisting in these regions, is naturally peaceful and timid. On the whole, however, his habits of life are artificial—more so than those of any domesticated animal. He is, therefore, in the conditions most favorable to variation. Again, man possesses more than merely animal instincts. His mental powers permit him to devise means of locomotion, of protection, of subsistence, far superior to those of any mere animal; and his dominant will, insatiable in its desires, bends the bodily frame to uses and exposes it to external influences

more various than any inferior animal can dream of. Man is also more educable and plastic in his constitution than other animals, owing both to his being less hemmed in by unchanging instincts, and to his physical frame being less restricted in its adaptations. If a single species, he is also more widely distributed than any other; and there are even single races which exceed in their extent of distribution nearly all the inferior animals. Nor is there anything in his structure specially to limit him to plains, or hills, or forests, or coasts, or inland regions. All the causes which we can suppose likely to produce variation thus meet in man, who is himself the producer of most of the distinct races that we observe in the lower animals. If, therefore, we condescend to compare man with these creatures, it must be under protest that what we learn from them must be understood with reference to his greater capabilities.

2. The races of men are deficient in some of the essential characters of species. It is true that they are reproduced with considerable permanency; though a great many cases of spontaneous change, of atavism, or return to the character of progenitors, and of slow variation under changed conditions, have been recorded. But the most manifest deficiency in true specific characters is in the invariable shading-off of one race into another, and in the entire failure of those who maintain the distinction of species in the attempt accurately to define their number and limits. The characters run into each other in such a manner that no natural arrangement based on the whole can apparently be arrived at; and when one particular ground is taken, as color, or shape of skull, the so-called species have still no distinct limits; and all the arrangements formed differ from each other, and from the deductions of philology and history. Thus, from the division of Virey into two species, on the entirely arbitrary ground of facial angle, to that of Bory de St. Vincent into fifteen, we have a great number and variety of distinctions, all incapable

of zoological definition; or, if capable of definition, eminently unnatural. There are, in short, no missing links between the varieties of men corresponding to that which obtains between man and lower animals.

3. The races of men differ in those points in which the higher animals usually vary with the greatest facility. The physical characters chiefly relied on have been color, character of hair, and form of skull, together with diversities in stature and general proportion. These are precisely the points in which our domestic races are most prone to vary. The manner in which these characters differ in the races of men may be aptly illustrated by a few examples of the arrangements to which they lead.

Dr. Pickering, of the U. S. Exploring Expedition* — who does not, however, commit himself to any specific distinctions —has arranged the various races of men on the very simple and obvious ground of color. He obtains in this way four races—the White, the Brown, the Blackish-brown, the Black. The distinction is easy; but it divides races historically, philologically, and structurally alike; and unites those which, on other grounds, would be separated. The white race includes the Hamite Abyssinian, the Semitic Arabian, the Japhetic Greek. The Ethiopian or Berber is separated from the cognate Abyssinian, and the dark Hindoo from the paler races speaking like him tongues allied to the Sanscrit. The Papuan, on the other hand, takes his place with the Hindoo; while the allied Australian must be content to rank with the Negro; and the Hottentot is promoted to a place beside the Malay. It is unnecessary to pursue any farther the arrangement of this painstaking and conscientious inquirer. It conclusively demonstrates that the color of the varieties of the human race must be arbitrary and accidental, and altogether independent of unity or diversity of origin.

---

* "The Races of Men," etc. Boston, 1848.

Some use has been made, by the advocates of diversity of species, of the quality of the hair in the different races. That of the Negro is said to be flat in its cross section—in this respect approaching to wool; that of the European is oval; and that of the Mongolian and American round.[*] The subject has as yet been very imperfectly investigated; but its indications point to no greater variety than that which occurs in many domesticated animals—as, for instance, the hog and sheep. Nay, Dr. Carpenter states[†]—and the writer has satisfied himself of the fact by his own observation—that it does not exceed the differences in the hair from different parts of the body of the same individual. The human hair, like that of mammals in general, consists of three tissues: an outer cortical layer, marked by transverse striæ, having in man the aspect of delicate lines, but in many other animals assuming the character of distinct joints or prominent serrations; a layer of elongated, fibrous cells, to which the hair owes most of its tenacity; and an inner cylinder of rounded cells. In the proportionate development of these several parts, in the quantity of coloring matter present, and in the transverse section, the human hair differs very considerably in different parts of the body. It also differs very markedly in individuals of different complexions. Similar but not greater differences obtain in the hair of the scalp in different races; but the flatness of the Negro's hair connects itself inseparably with the oval of the hair of the ordinary European, and this with the round observed in some other races. It generally holds that curled and frizzled hair is flatter than that which is lank and straight; but this is not constant, for I have found that the waved or frizzled hair of the New Hebrideans, intermediate apparently between the Polynesians and Papuans, is nearly circular in outline, and differs from European hair mainly in the greater

---

[*] Browne, of Philadelphia, quoted by Kneeland and others.
[†] Todd's Cyclopædia, art. "Varieties of Man."

development of the fibrous structure and the intensity of the color. Large series of comparisons are required; but those already made point to variation rather than specific difference. Some facts also appear to indicate very marked differences as occurring in the same race from constant exposure or habitual covering; and also the occasional appearance of the most abnormal forms, without apparent cause, in individuals. The differences depending on greater or less abundance or vigor of growth of the hair are obviously altogether trivial, when compared with such examples as the hairless dogs of Chili and hairless cattle of Brazil, or even with the differences in this respect observed in individuals of the same race of men.

Confessedly the most important differences of the races of men are those of the skeleton, in all parts of which variations of proportion occur, and are of course more or less communicated to the muscular investments. Of these, as they exist in the pelvis, limbs, etc., I need say nothing; for, manifest though they are, they all fall far within the limits of variation in familiar domestic animals, and also of hereditary malformation or defect of development occurring in the European nations, and only requiring isolation for its perpetuation as a race. The differences in the skull merit more attention, for it is in this and in its enclosed brain that man most markedly differs from the lower animals, as well as race from race. It is in the form rather than in the mere dimensions of the skull that we should look for specific differences; and here, adopting the vertical method of Blumenbach as the most characteristic and valuable, we find a greater or less antero-posterior diameter — a greater or less development of the jaws and bones of the face. The skull of the normal European, or Caucasian of Cuvier, is round oval; and the jaws and cheek-bones project little beyond its anterior margin, when viewed from above. The skull of the Mongolian of Cuvier is nearly round, and the cheek-bones and jaws project much more strongly in front and at the sides. The Negro skull is lengthened from back to front; the jaws project strongly, or are prognathous; but the cheek-bones are little prominent. For the extremes of these varieties, Retzius proposed the names of brachy-kephalic or short-headed, and dolicho-kephalic or long-headed, which have come into general use. The differences indicated by these terms are of great interest, as distinctive marks of many of the unmixed races of men; but, when pushed to extremes, lead to very incorrect generalizations—as Professor D. Wilson has well shown in his paper on the supposed uniformity of type in the American races— a doctrine which he fully refutes by showing that within a very narrow geographical range this primitive and unmixed race presents very great differences of cranial form.* Exclusive of idiots, artificially compressed heads, and deformities, the differences between the brachy-kephalic and dolicho-kephalic heads range from equality in the parietal and longitudinal diameter to the proportion of about 14 to 24. As stated by some ethnologists, these differences appear quite characteristic and distinct; but, so soon as we attempt any minute discrimination, all confidence in them as specific characters disappears. In our ordinary European races similar differences, and nearly as extensive, occur. The dolicho-kephalic head is really only an immature form perpetuated; and appears not only in the Negro, but in the Esquimau, and in certain ancient and modern Celtic races. The brachy-kephalic head, in like manner, is characteristic of certain tribes and portions of tribes of Americans, but not of all; of many northern Asiatic nations; of certain Celtic and Scandinavian tribes; and often appears in the modern European races as an occasional character. Farther, as Retzius has well shown, the long heads and prominent jaws are not always associated with each other; and his classification is really the testimony of an able observer

---

* "Prehistoric Man."

against the value of these characters. He shows that the Celtic and Germanic races (in part) have long heads and straight jaws; while the Negroes, Australians, Oceanians, Caribs, Greenlanders, etc., have long heads and prominent jaws. The Laplanders, Finns, Turks, Sclaves, Persians, etc., have short heads and straight jaws; while the Tartars, Mongolians, Incas, Malays, Papuans, etc., have short heads and prominent jaws.

Another defect in the argument often based on the diverse forms of heads is its want of acknowledgment of the ascertained and popularly known fact that these forms in different tribes or individuals of the same race are markedly influenced by culture and habits of life. In all races ignorance and debasement tend to induce a prognathous form, while culture tends to the elevation of the nasal bones, to an orthognathous condition of the jaws, and to an elevation and expansion of the cranium.*

Again, no adequate allowance has been made in the case of these forms of skull for the influence of modes of nurture in infancy. Dr. Morton, observing that the brachy-kephalic American skull was often unequal sided, and the occiput much flattened, suggests that this is "an exaggeration of the natural form produced by the pressure of the cradle-board in common use among the American natives." Dr. Wilson has noticed the same unsymmetrical character in brachy-kephalic skulls in British barrows, and has suspected some artificial agency in infancy; and says, in reference to the American instances, "I think it extremely probable that further investigation will tend to the conclusion that the vertical or flattened occiput, instead of being a typical characteristic, pertains entirely to the class of artificial modifications of the natural cranium familiar to the American ethnologist."

While the points in which the races of men vary are those

---

* Carpenter in Todd's Cyclopædia.

in which lower animals are most liable to undergo change, the several races display a remarkable constancy in those which are usually less variable. Prichard and Carpenter have well shown this in relation to physiological points, as, for instance, the age of arriving at maturity, the average and extreme duration of life, and the several periods connected with reproduction. The coincidence in these points alone is by many eminent physiologists justly regarded as sufficient evidence of the unity of the species.

4. It may also be affirmed, in relation to the varieties of man, that they do not exceed in amount or extent those observed in the lower animals. If with Frederick Cuvier, Dr. Carpenter, and many other naturalists, we regard the dog as a single species, descended in all probability from the wolf, we can have no hesitation in concluding that this animal far exceeds man in variability.* But this is denied by many, not without some show of reason ; and we may, therefore, select some animal respecting which little doubt can be entertained. Perhaps the best example is the common hog (*Sus scrofa*), an undoubted descendant of the wild boar, and a creature especially suitable for comparison with man, inasmuch as its possible range of food is very much the same with his, which is not the case with any other of our domesticated animals ; and as its headquarters as a species are in the same regions which have supported the greatest and oldest known communities of men. We may exclude from our comparison the Chinese hog, by some regarded as a distinct species (*Sus Indicus*), though no wild original is known, and it breeds freely with the common hog. The color of the domestic hog varies, like that of man, from white to black ; and in the black hog the skin as well as the hair partakes of the dark color. The

---

* For an interesting inquiry into the origin of the dog, see the article in Todd's Cyclopædia already referred to ; and the subject is fully discussed by Darwin, who leans to the theory of the diversity of origin in dogs.

abundance and quality of the hair vary extremely ; the stature and form are equally variable, much more so than in man.   Blumenbach long ago remarked that the difference between the skull of the ordinary domestic hog and that of the wild boar is quite equal to that observed between the Negro and European skulls.   Darwin shows that it is much greater, and illustrates this by an amusing pair of portraits. The breeds of swine even differ in directions altogether unparalleled in man.   For instance, both in America and Europe solid-hoofed swine have originated and become a permanent variety ; and there is said to be another variety with five toes.*   These are the more remarkable, because, in the American instances, there can be no doubt that it is the common hog which has assumed these abnormal forms.

5. All varieties or races of men intermix freely, in a manner which strongly indicates specific unity.   We hold here, as already stated, that no good case of a permanent race arising from intermixture of distinct species of the lower animals has been adduced ; but there is another fact in relation to this subject which the advocates of specific diversity would do well to study.   Even in varieties of those domestic animals which are certainly specifically identical, as the hog, the sheep, the ox—although crosses between the varieties may easily be produced—they are not readily maintained, and sometimes tend to die out.   What are called good crosses lead to improved energy, and continual breeding in and in of the same variety leads to degeneracy and decay ; but, on the other hand, crosses of certain varieties are proved by experience to be of weakly and unproductive quality ; and every practical book on cattle contains remarks on the difficulty of keeping up crosses without intermixture with one of the pure breeds.   It would thus appear that very unlike varieties of the same species display in this respect, in an imperfect man-

---

* Prichard, Bachman, Cabell.

ner, the peculiarities of distinct species. It is on this principle that I would in part account for some of the exceptional
facts which occur in mixed races of men.

What, then, are the facts in the case of man? In producing crosses of distinct species, as in the case of the horse and
ass, breeders are obliged to resort to expedients to overcome
the natural repugnance to such intermixture. In the case of
even the most extreme varieties of man, if such repugnance
exists, it is voluntarily overcome, as the slave population of
America testifies abundantly. By far the greater part of the
intermixtures of races of men tend to increase of vital energy
and vigor, as in the case of judicious crosses of some domestic animals. Where a different result occurs, we usually find
sufficient secondary causes to account for it. I shall refer to
but one such case—that of the half-breed American Indian.
In so far as I have had opportunities of observation or inquiry, these people are prolific, much more so than the unmixed Indian. They are also energetic, and often highly
intellectual; but they are of delicate constitution, especially
liable to scrofulous diseases, and therefore not long-lived.
Now this is precisely the result which often occurs in domestic animals, where a highly cultivated race is bred with one
that is of ruder character and training; and it very probably
results from the circumstance that the progeny may inherit
too much of the delicacy of the one parent to endure the
hardships congenial to the other ; or, on the other hand, too
much of the wild nature of the ruder parent to subsist under
the more delicate nurture of the more cultivated. This difficulty does not apply to the intermixture of the Negro and
the European, though between the pure races this is a cross
too abrupt to be likely to be in the first instance successful.

6. The races of man may have originated in the same manner with the breeds of our domesticated animals. There are
many facts which render it probable that they did originate

in this way. Take color, for instance. The fair varieties of man occur only in the northern temperate zone, and chiefly in the equable climates of that zone. In extreme climates, even when cold, dusky and yellow colors appear. The black and blackish-brown colors are confined to the inter-tropical regions, and appear in such portions of all the great races of mankind as have been long domiciled there. Diet and degree of exposure have also evidently very much to do with form, stature, and color. The deer-eating Chippe-wayan of certain districts of North America is a better devel-oped man than his compatriots who subsist principally on rabbits and such meaner fare ; and excess of carbonaceous food, and deficiency of perspiration or of combustion in the lungs, appear everywhere to darken the skin.* The Negro type in its extreme form is peculiar to low and humid river valleys of tropical Africa. In Australasia similar characters appear in men of a very different race in similar circum-stances. The Mongolian type reappears in South Africa. The Esquimau is like the Fuegian. The American Indian, both of South and North America, resembles the Mongol ; but in several of the middle regions of the American con-tinent men appear who approximate to the Malay. Every-where and in all races coarse features and deviations from the oval form of skull are observed in rude populations. Where men have sunk into a child-like simplicity, the elon-gated forms prevail. Where they have become carnivorous, aggressive, and actively barbarous, the brachy-kephalic forms abound. These and many other considerations tend to the conclusion that these varieties are inseparably connected with external conditions. It may still be asked—Were not the races created as they are, with especial reference to these

* A curious note, by Dr. John Rae, on the change of complexion in the Sandwich Islanders, consequent on the introduction of clothing, may be found in the "Montreal Medical Chronicle," 1856, and the "Canadian Journal" for the same year.

conditions? I answer no—because the differences are of a character in every respect like those that appear in other true species as the results of influences from without.

Farther, not only have we varieties of man resulting from the slow operation of climatal and other conditions, but we have the sudden development of races. One remarkable instance may illustrate my meaning. It is the hairy family of Siam, described by Mr. Crawford and Mr. Yule.* The peculiarities here consisted of a fine silky coat of hair covering the face and less thickly the whole body, with at the same time the entire absence of the canine and molar teeth. The person in whom these characters originated was sent to Ava as a curiosity when five years old. He married at twenty-two, his wife being an ordinary Burmese woman. One of two children who survived infancy had all the characters of the father. This was a girl; and on her marriage the same characters reappeared in one of two boys constituting her family when seen by Mr. Yule. Here was a variety of a most extreme character, originating without apparent cause, and capable of propagation for three generations, even when crossed with the ordinary type. Had it originated in circumstances favorable to the preservation of its purity, it might have produced a tribe or nation of hairy men, with no teeth except incisors. Such a tribe would, with some ethnologists, have constituted a new and very distinct species; and any one who had suggested the possibility of its having originated within a few generations as a variety would have been laughed at for his credulity. It is unnecessary to cite any further instances. I merely wish to insist on the necessity of a rigid comparison of the variations which appear in man, either suddenly or in a slow or secular manner, with the characters of the so-called races or species.

7. If we turn from the merely physical constitution of man,

---

* Latham's " Descriptive Ethnology."

T

and inquire as to his psychical and spiritual endowments, it would be easy to show, as Dr. Carpenter and others have done, in opposition to Darwin, that on the one hand an impassable barrier separates man from the lower animals, and that on the other there is an essential unity among the races of men. But this subject I have discussed fully in the concluding chapters of my " Story of the Earth."

If man is thus so very variable, and if many of his leading varieties have existed for a very long time, does not the fact that we have but one species afford very strong evidence that species change only within fixed limits, and do not pass over into new specific types. Viewed in this way, variability within the specific limits becomes in itself one of the strongest arguments against the doctrine of descent with modification as a mode of origination of new species.

Let us now add to all this the farther consideration, so well illustrated in the " Reliquiæ Aquitanicæ " of Christy and Lartet, that the oldest-known men of the caves and gravels may be placed in one of the varieties, and this the most widely distributed, of modern man, and we have a further argument which tells most strongly against the assumption either of the extreme antiquity or of the unlimited variability of the human species.

INDEX.

THE END.

www.ingramcontent.com/pod-product-compliance
Lightning Source LLC
Chambersburg PA
CBHW021344210326
41599CB00011B/749